Teaching Mathematical Reasoning in Secondary School Classrooms

Karin Brodie

Teaching Mathematical Reasoning in Secondary School Classrooms

With Contributions by

Kurt Coetzee
Lorraine Lauf
Stephen Modau
Nico Molefe
Romulus O'Brien

 Springer

Karin Brodie
School of Education
University of the Witwatersrand
Johannesburg
South Africa
karin.brodie@wits.ac.za

ISBN 978-0-387-09741-1   e-ISBN 978-0-387-09742-8
DOI 10.1007/978-0-387-09742-8
Springer New York Dordrecht Heidelberg London

Library of Congress Control Number: 2009935695

Printed on acid-free paper

Springer is part of Springer Science+Business Media (www.springer.com)

Foreword

The Road to Reasoning

The teachers in this book share a worthy and courageous mission. They have all set out to provide children with one of the most important educational experiences it is possible to have – a form of mathematics teaching that is based upon sense making and discussion, rather than submission and silence. Mathematical "reasoning" is what mathematicians do – it involves forming and communicating a path between one idea or concept and the next. When students form these paths they come to enjoy mathematics, understand the reasons why ideas work, and develop a connected and powerful form of knowledge. When students do not engage in reasoning, they often do not know that there are paths between different ideas in mathematics and they come to believe, dangerously, that mathematics is a set of isolated facts and methods that need to be remembered. I have visited hundreds of classrooms across the world in which students have been required to work in silence on maths questions, never talking about the ideas or forming links and connections between ideas; most of these students come to dislike mathematics and drop the subject as soon as they can. Such students are not only being denied the opportunity to learn in the most helpful way, but they are denied access to real, living mathematics.

The teachers in this book, through their work with Karin Brodie, the author, learned about the value of mathematical reasoning and set out to teach students to engage in this valuable act. This book shares their important journey and provides the world with new lenses for considering the teaching acts that were involved, as well as the challenges and obstacles that stood in their way. For whilst we know the importance of reasoning to children's mathematical futures it would be dishonest to pretend that teaching approaches that invite students to communicate their mathematical thoughts and make connections between ideas are easy or well understood. We have reached an advanced stage in the development of education and yet, incredibly, we are still relatively uninformed about the ways teachers of mathematics can teach students to reason, which is part of the reason this book is so valuable and could be a wonderful resource for many.

When Deborah Ball, in the United States, then an elementary teacher of mathematics, now a university dean, released a videotape of her teaching 7- and 8-year olds to reason about odd and even numbers, the world was shocked to witness a boy named Shea propose a new way of classifying odd numbers. His numbers – those that can be grouped into even numbers of pairs of twos – came to be known as "Shea numbers". The rich conversations in which the young children engaged in the mathematics class that appeared on tape, seemed to unfold effortlessly, although in reality they were expertly choreographed by the teacher. Deborah Ball has offered records of her teaching decisions and actions, which have been read by scores of people worldwide, including the teachers who write in this book. She was one of the first teachers to offer such valuable records and analyses. This book adds to the small but important collection of teachers who have engaged students in mathematical reasoning and documented and unpacked the important teaching acts that took place.

But what makes a record of teaching useful and worthwhile? Every act of teaching, with a classroom full of children and their many thoughts and actions, is extremely complex, and descriptions of a class in action can remain highly contextualized and difficult for others to learn from. A teacher may record thoughts and moves without communicating them in such a way that they are useful for other teachers, educators, and analysts. The art in producing a record that is powerful and valuable for others comes partly from having important teaching experiences to talk about and partly from having a way of raising the individual acts to a higher and more generalizable level that other teachers can learn from. This is where the combination of the reports of the teachers who engaged students in reasoning, and the theoretical lenses applied by Karin, are so generative and fruitful for the rest of the world to learn from. When a new idea and teaching act is connected with a theory of learning, the result can be very powerful indeed.

An example of the way a teaching act can be named and made more general is the case of a set of interactions that has become known as IRE. These describe a common teaching situation when a teacher initiates something (I), elicits a response from a student (R), and then evaluates the response (E). Researchers found that the majority of the interactions that take place in classrooms follow the IRE response pattern and they gave it a particular classification. Since that initial classification IRE has been used by scores of researchers and analysts over many years and has proved extremely useful in the advancement of teaching. Yet teaching classifications such as IRE are rare and the field of mathematics education has not benefitted from a similar mapping and classification of the teaching interactions that take place when students are taught to reason about mathematics. This book provides such a mapping.

Karin notes that a reasoning approach to mathematics involves a change in authority. Students no longer need to look to teachers or textbooks to know if they are moving in the right directions in mathematics, as they have learned a set of reasons and connections that they can refer back to, evaluating their own thoughts and ideas. This may seem as though the authority is shifting from the teacher to the students and this is partly true, but it is important to note that the authority is also shifting from the teacher to the domain of mathematics itself. Students no longer need to refer to teachers to evaluate their mathematical thoughts, because they can refer to the domain of mathematics, to consider whether they have followed the correct connections and paths. This is just one way in which reasoning as an act brings classrooms closer to real and living mathematics. In addition, we now have evidence that when students receive opportunities to discuss mathematics and express their own thoughts, they become more open-minded as they learn to be appreciative and respectful of other people's ideas. Mathematical reasoning encourages respect, responsibility, and a personal empowerment that has long been missing in mathematics classrooms. Karin starts this book by quoting the goals of the new South African curriculum – to heal the divisions of the past and build a human rights culture. Mathematics, the subject so many believe to be abstract and removed from such responsibilities, has a key role to play in promoting such a culture, in South Africa and beyond. This book communicates the way that mathematics can provide this valuable contribution and the important work of teachers in doing so. I hope you enjoy it and use it as both inspiration and resource.

Jo Boaler   The University of Sussex

Contents

List of Tables

Introduction to Part 1

Over the past 10 years, South Africa has introduced a new curriculum in all subjects at all levels of schooling. The curriculum was inspired by the end of apartheid and informed by curriculum developments and visions of reform in many other countries. The new curriculum has an impressive set of goals for individual learners and society, including healing the divisions of the past, building a human rights culture, and developing skilled and knowledgeable citizens who can contribute to and benefit from a growing economy and a participatory democracy (Department of Education 2003).

The new curriculum posits a very different view of mathematical knowledge from that of previous curricula. Mathematics is seen as both conceptual and practical; abstract and applied. The curriculum argues for conceptual understanding of mathematical ideas, skill in performing mathematical calculations, and the ability to relate mathematical concepts to other subjects and to real-world applications. Mathematical concepts and skills are developed and linked by "creative and logical reasoning" and "rigorous logical thinking" (Department of Education 2003, p. 9). The new curriculum puts mathematical reasoning firmly on the agenda arguing "competence in mathematical process skills such as investigating, generalizing and proving is more important than the acquisition of content knowledge for its own sake" (Department of Education 2003, p. 9). A view of mathematics as a combination of conceptual depth, flexible skills, and mathematical reasoning resonates with curriculum developments elsewhere in the world. Some proponents of reform mathematics argue that reasoning should be taught alongside many of the basic facts and concepts of mathematics and some even argue that mathematical reasoning is in itself a "basic" mathematical skill (Ball and Bass 2003).

Worldwide, too many learners struggle with mathematics, fail mathematics, and hate mathematics. Moreover, when the factors of race and class are considered, it is clear that a disproportionate number of black and economically disadvantaged learners do not achieve success in mathematics and do not believe that they can do mathematics (Association for Mathematics Education of South Africa 2000; Department of Education 2001; Moses and Cobb 2001; Secada 1992). The proponents of reform curricula propose that new approaches to mathematics and to teaching mathematics will make mathematics more accessible, enjoyable, and inspiring for many more learners, enable more learners to be successful in mathematics and will begin to close achievement gaps between rich and poor and black and white learners.

There has been much debate as to whether current mathematics reforms can be a mechanism for ensuring more equitable participation and achievement in mathematics (see Brodie 2006 for a summary of these debates). Empirical evidence in well-resourced countries is beginning to show that curriculum reforms do mitigate achievement gaps between marginalized and other learners (Boaler 1997; Hayes et al. 2006; Kitchen et al. 2007; Schoenfeld 2002). However, the evidence also suggests an important caveat. The implementation of the new curriculum ideas is not widespread and is inequitably distributed, tending to be found in more well-resourced schools and countries, hence further disadvantaging poor learners (Kitchen et al. 2007). There are two important implications of these findings. First, in many contexts, lack of resources, including big classes and few materials, teacher confidence and knowledge, and support for teachers, can be major barriers to developing new ways of teaching (Tabulawa 1998; Tatto 1999). If reforms are successful in promoting equity and if they are not taken up in less-resourced countries and schools, then existing divides between rich and poor will be exacerbated. Second, it is not only a lack of resources that creates barriers to reform practice. In fact, research in many well-resourced contexts suggests that most teachers struggle to take up reform practices in substantive ways (Fraivillig et al. 1999; Hayes et al. 2006; Hufferd-Ackles et al. 2004; Lavi and Shriki 2008; Nolan 2008). This suggests that reasons for the difficulties that teachers experience with reform curricula cannot be found in resources alone. Something deeper is at play here.

The research on teacher change in the context of reform is based on the assumption, shared in this book, that teachers are key to achieving the visions of better mathematics learning for more learners. This research can be divided into two main categories. The first describes models of exemplary reform teaching, making the claim that such teaching is possible, albeit with many challenges, illuminating different approaches to reform teaching and showing how the challenges can be overcome. Such cases come mainly from well-resourced contexts, further adding to the concern that such teaching is only possible in these contexts (Boaler 1997; Boaler and Humphreys 2005; Chazan and Ball 1999; Hayes et al. 2006; Heaton 2000; Lampert 2001; Staples 2007). The second set of research, which includes some of my own work, argues that teachers do not substantially shift their practices, even after extensive pre-service or in-service courses (Brodie et al. 2002; Fraivillig et al. 1999; Lavi and Shriki 2008; Nolan 2008; Tatto 1999). These two sets of findings tend to dichotomize the field, suggesting an ideal "reform vision", which unfortunately only a few can attain. A disturbing consequence of dichotomizing the field in this way is that teachers are often blamed for not being able to implement visions of reform because they do not live up to the ideals.

A far more promising line of research takes the middle road, presenting more textured descriptions of points of difficulty for teaching and when, how and why teaching in reform-oriented ways breaks down (Gamoran Sherin 2002). In my own work with colleagues and students, we have shown that some aspects of reform practice are easier for teachers to work with, for example selecting tasks of higher cognitive demand (Modau and Brodie 2008) whereas others are more difficult, for example interacting with students while maintaining the level of task demand (Jina and Brodie 2008; Modau and Brodie 2008; Stein et al. 1996, 2000). We have also argued that adopting reforms requires teachers to coordinate a range of new practices and to think about their current practices in new ways. Such coordination is an immense task and means that teachers' taken-for-granted practices might break down in the face of the new curriculum practices (Slonimsky and Brodie 2006). It is thus highly likely that teachers attempting to work with reforms may resort to traditional practices, more or less deliberately (Brodie 2007c).

This book aims to contribute to this third emerging strand of research and to give substance to a number of claims that such research can make. The first is that teachers' difficulties in working with new curricula need to be taken seriously, because, as we show in this book, mathematical reasoning is challenging to learn and to teach. For teachers who learned mathematics and learned to teach mathematics in traditional ways, the challenges are enormous. However, this does not mean that teachers cannot begin to work towards teaching mathematical reasoning in ways suggested by the reforms. But it does mean that researchers and teacher educators need to find ways to capture teachers' successes and challenges in ways that can help teachers to move forward. The successes might be small and the challenges might be large, but we need to find ways to show where and how teachers are shifting and what the next steps for progress might be. It is our experience, some of which we hope to share in this book, that teachers who do take risks and embark on the journey of learning to teach in new ways, have experienced both the exhilaration of success, when learners actually do begin to reason with each other and their teachers, and the extreme challenges of sustaining the practice with all learners, particularly given overcrowded curricula and high stakes tests and examinations.

So this book explores some of the successes and challenges faced by a group of South African teachers who worked to develop mathematical reasoning among their learners. In doing so, it explores what it means to teach mathematical reasoning in secondary school mathematics classrooms, addressing some important questions like what mathematical reasoning means; how can mathematical reasoning be taught; how teaching mathematical reasoning differs from more conventional mathematics teaching; and the demands that teaching mathematical reasoning makes on teachers and learners. A number of chapters address these questions from the perspective of the teachers analysing their own practice, and others address the questions from the perspective of an academic researcher, analysing the teachers' practices.

The book is the result of an ongoing collaboration between five teachers and an academic. We came together with a joint interest in promoting mathematical reasoning in South African classrooms. We share a passion for improving the experiences of learners in mathematics classrooms, and a belief that working in reform-oriented ways can do this. At the same time, we experience the real constraints of classrooms and are inspired by the need to find ways to work with contextual realities to support mathematical reasoning. Although this book is set in a

South African context, it has strong implications for other contexts. This work was conducted in a range of differently resourced classrooms, from very poorly resourced to very well resourced, and so can illuminate the teaching of mathematical reasoning in relation to contextual differences and speak to readers working in a wide range of countries and schools.

Each teacher conducted her/his study as the research component of an Honours programme in mathematics education, in which they were all enrolled. At the time, I was engaged in my doctoral research, and worked with the teachers as subjects of my research and as advisor on their research projects. Each teacher worked on a subset of her/his data from the larger data corpus, which I collected. We worked closely together as a group and the results of each study informed the others. The dilemmas, strengths, and challenges of such a collaboration have been discussed in detail elsewhere (Brodie 2005; Brodie et al. 2005).

Outline of the Book

This book comprises three parts. Part 1, consisting of Chaps. 1 and 2, sets the context of the work within the literature on teaching and learning mathematical reasoning and describes the school contexts in which the research took place. In Chap. 1, I review the literature and develop the concepts of mathematical reasoning, learning mathematical reasoning and teaching mathematical reasoning, that informed the work of the project and the subsequent chapters of this book. In Chap. 2, I discuss the differently resourced contexts in which the teachers worked. I argue that the learners' knowledge in the different classrooms forms a substantive part of the teaching context. Through the development of this work, we came to see learners' knowledge as a key resource for teaching mathematical reasoning. I also discuss the tasks that the teachers developed to teach mathematical reasoning and which they refer to in their chapters.

Part 2, consisting of Chaps. 3–7, describes the studies that the teachers conducted. Each teacher researched an aspect of her/his practice, trying to understand more deeply the challenges and successes that she/he experienced. These chapters were informed by some of the literature discussed in Chap. 1 and each teacher chose particular parts of the literature to work with as a conceptual framework. In Chap. 3, we take a close look at how a set of tasks supported learners' reasoning and in Chap. 4 we explore how collaborative, whole-class discussion supported both individual and group learning. In Chap. 5, we describe a set of practices, which the teacher developed to enable learners to reason with each other and show how his learners appropriated these practices to help with their thinking and reasoning. In Chap. 6, we show how the teacher supported the development of mathematical proficiency among her learners and in Chap. 7, we focus on the development of justification among learners. Taken together, these chapters provide a rich account of challenges and successes in teaching mathematical reasoning, and what is possible to achieve even in very difficult circumstances. They deal with a number of key aspects in teaching mathematical reasoning, namely, what tasks might be useful to elicit learners' reasoning and how are they best implemented in the classroom, how different tasks support learners' mathematical reasoning in different ways, and how classroom interaction helps to support the development of mathematical reasoning.

Part 3, consisting of Chaps. 8–11, comprises an overview of the practices in the five classrooms, drawing on my doctoral research. Chapters 8 and 9 look across all five classrooms and develop categories for talking about learner contributions and teacher moves, as the beginnings of a language of description for reform-oriented teaching. These two chapters together suggest a trajectory for the emergence of learner contributions and teachers' responses that promote mathematical reasoning. The argument is that by finding more specific ways to talk about how teachers and learners interact in classrooms, we can find ways to help teachers move forward in engaging learners' mathematical reasoning. Chapter 10 focuses in more depth on two teachers and the dilemmas they experienced in teaching mathematical reasoning. Chapter 11 focuses even more closely on one classroom and explores the resistance of the learners to their new experience of learning mathematical reasoning. Such resistance is often reported by teachers but not often explored. In this chapter I suggest that resistance is an important aspect of the new methods of teaching and suggest ways of managing it. The overarching argument of Part 3 is that every success in reform pedagogy produces new challenges, a range of learner contributions to respond to, new dilemmas in relation to these contributions, and possible learner resistance. Teachers, teacher–educators, and researchers cannot ignore these challenges; we have to find ways to talk about them as a normal part of learning to teach in new ways, and the challenges that change brings.

Because teachers have been involved in the work of this book and in writing parts of it, we hope that the work will speak to teachers, and to teacher educators and researchers who are trying to work in and with new curriculum developments. We have written this book for both teachers and researchers because we strongly believe that teachers and researchers can and should speak to each other in many ways, including through books such as this one. At the same time, we also realize that teaching and research are distinct practices, with their own discourses. This has caused some discomfort in the writing of this book in that we have had to continually consider two different audiences. We have resolved this by writing some chapters in a more "academic" tone and others somewhat more colloquially. In doing this, we have made sure to keep our research focus strong and rigorous throughout the book. In particular Chap. 1 sets out the academic field of teaching and learning mathematical reasoning and may not be the best part of the book for teachers to start reading. The case studies in Chaps. 3–8 are structured to form part of the ongoing narrative of the book as a whole but can also be read individually. In these chapters, key parts of the literature have been revisited for the purposes of the particular case study, and although this has meant some repetition of key ideas, we believe it will help readers to see these ideas working in different contexts. Chapters 8 and 9 develop a language of description, which might be more appealing to researchers than teachers. Chapters 10 and 11 deal with particular issues in two of the classrooms, and although building on the language of Chaps. 8 and 9, can also be read on their own.

In our work together, practice has spoken closely to research, and so we hope that the research described here will find ways to speak to practice. We believe that the work in this book will provide useful models for other teachers wanting to teach mathematical reasoning, for teachers wanting to research their own practice, and for teacher–educators and researchers wanting to develop and analyse the teaching of mathematical reasoning. We undertook a journey together in which we learned a tremendous amount. We hope to convey some of it in this book and also inspire others to embark on similar journeys.

Curriculum reform has become a global movement over the past 30 years. Similarities among the South African and other mathematics curricula will be discussed further in this chapter and the book.

Terminology used across contexts to refer to new curricula is different. In this book we use interchangeably the terms "new curriculum" which applies to South Africa and other countries, which have national curricula, and "reform" which is used predominantly in the United States. In both cases we include the enacted curriculum, i.e. teaching and learning in classrooms.

In the South African Higher Education system an Honours degree follows a 3-year undergraduate degree and a professional teaching qualification and is necessary for entry into Masters.

Chapter 1
Teaching Mathematical Reasoning: A Challenging Task

The Centrality of Mathematical Reasoning in Mathematics Education

When we "reason", we develop lines of thinking or argument, which might serve a number of purposes – to convince others or ourselves of a particular claim; to solve a problem; or to integrate a number of ideas into a more coherent whole. Two processes are important to reasoning – first, that the different steps or moves in the line of reasoning are connected with each other (not necessarily analytically or deductively); and second, that these links are somehow "reasoned", there are reasons why one move follows another and how a number of moves come together to form an argument or to solve a problem (Ball and Bass 2003). Brousseau and Gibel (2005) point out that these reasons are only considered to be reasonable when they relate to the constraints of the problem or the knowledge under consideration. An appeal to authority, for example to what a teacher or textbook says, does not count as a reason for a productive argument.

The product of a reasoning process is a text, either spoken or written (Douek 2005), which presents warrants for a conclusion that is acceptable within the community that is producing the argument (Krummheuer 1995). An individual can reason, or a group of people can reason together, co-producing the line of argument[1]. Mathematical reasoning assumes mathematical communication (Ball and Bass 2003; Douek 2005; Krummheuer 1995). Communication is an integral part of the process of reasoning, both for an individual working with previously produced texts to produce a new one, and for groups working together to produce an argument. The texts or products of reasoning have, as their main purpose, to communicate reasoning.

Mathematical reasoning is reasoning about and with the objects of mathematics. However, the relationship between mathematical reasoning and mathematics is not obvious (Steen 1999), and the processes involved in mathematical reasoning need

---

[1] Social perspectives on learning and thinking would argue that even an individual reasoning, seemingly on her own, is in fact in dialogue with others, co-producing an argument, with an imagined audience, with ideas from others, and in a social and historical context (see below).

K. Brodie, *Teaching Mathematical Reasoning in Secondary School Classrooms*,
DOI 10.1007/978-0-387-09742-8_1, © Springer Science+Business Media, LLC 2010

some elaboration. For Ball and Bass (2003) reasoning is a "basic skill" (p. 28) of mathematics and is necessary for a number of purposes – to understand mathematical concepts, to use mathematical ideas and procedures flexibly, and to reconstruct once understood, but forgotten mathematical knowledge. Kilpatrick et al. (2001) define a notion of mathematical proficiency which requires five intertwined and mutually influential strands – conceptual understanding, which entails comprehension of mathematical concepts, operations, and relations; procedural fluency, involving skill in carrying out procedures flexibly, accurately, efficiently, and appropriately; strategic competence, which is the ability to formulate, represent, and solve mathematical problems; adaptive reasoning, which is the capacity for logical thought, reflection, explanation, and justification; and productive disposition, an orientation to seeing mathematics as sensible, useful, worthwhile, and reasonable, and that anyone can reason to make sense of mathematical ideas[2]. For Kilpatrick et al. (2001), although all the strands are important and mutually influential, "adaptive reasoning is the glue that holds everything together" (p. 129) in that it allows for concepts and procedures to connect together in sensible ways, suggests possibilities for problem solving, and allows for disagreements to be settled in reasoned ways. Central to adaptive reasoning is the justification of claims and development of arguments.

This view of mathematical proficiency has informed all of the work in this book. Most directly, in Chap. 6 we reflect on one teacher's attempt to teach the five strands in a holistic way. The teacher found that she devoted most of the time to conceptual understanding rather than procedural fluency, which is traditionally the norm in mathematics classrooms (Kilpatrick et al. 2001; Schoenfeld 1988; Stigler and Hiebert 1999). However, she was concerned that she devoted less time to strategic competence and adaptive reasoning. She also found that more than half of the learners in her class showed evidence of all five strands in their written work. In Chaps. 5 and 7 we focus on the strand of adaptive reasoning and show how two teachers supported learners to reason adaptively.

Justifying and Generalizing

The literature suggests that there are two key practices involved in mathematical reasoning – justifying and generalizing – and other mathematical practices such as symbolizing, representing, and communicating, are key in supporting these (Ball 2003; Ball and Bass 2003; Davis and Maher 1997; Triandafillidis and Potari 2005). For Kilpatrick et al. justifying is a key element of adaptive reasoning and to justify means "to provide sufficient reason for" (p. 130). They argue "students need to be able to justify and explain ideas in order to make their reasoning clear, hone their

---

[2] I note here that Kilpatrick et al.'s work is an extension of the more usual distinctions of conceptual and procedural understandings of mathematics (Hiebert and Lefevre, 1986).

reasoning skills and improve their conceptual understanding" (p. 130). For Ball and Bass, "unjustified knowledge is unreasoned and, hence, easily becomes unreasonable" (p. 29). Justification is a key mathematical practice that allows mathematicians and mathematics teachers and learners to make connections between different ideas and parts of an argument, to provide warrant for claims and conjectures, to settle disputes, and to develop new mathematical ideas.

For Russell (1999), mathematical reasoning is "essentially about the development, justification and use of mathematical generalizations" (p. 1). These generalizations create an interconnected web of mathematical knowledge – conceptual understanding in Kilpatrick et al.'s terms. For Russell, "seeing mathematics as a web of interrelated ideas is both a result of an emphasis on mathematical reasoning and a foundation for reasoning further" (p. 5). Creating generalizations also enables problem solving, as generalizations support learners to see the underlying structure of the problem and the bigger class of problems or ideas that it instantiates (Brousseau and Gibel 2005; Kilpatrick et al. 2001; Russell 1999). Russell also introduces a notion of "mathematical memory", which is a memory of fundamental mathematical relationships, rather than of isolated facts. This kind of memory is what allows mathematical knowers to reconstruct, in a reasoned way, mathematical concepts, procedures, and principles that they might have forgotten (Ball and Bass 2003; Brousseau and Gibel 2005). It also supports sense making and insight in mathematics, and creates the conditions for solving problems.

In Chap. 7, we directly address the challenges that a teacher faced in supporting his learners to justify their thinking. The vast majority of learners in his class were not able to answer the question: "can x^2+1 be less than zero, when x is a real number", with appropriate justifications. We show how the teacher worked through a number of different contributions from learners, ranging from incorrect justifications through those that were partially correct, to one that was completely correct, asking them to discuss and communicate their reasoning. Even though his learners had very weak mathematical knowledge, they were, with a lot of help from the teacher, able to contribute and to help each other develop better justifications. In each of the other teachers' chapters, we see examples of learners' successes and challenges as they work to justify, explain, and generalize their ideas.

The Role of Proof in Mathematical Reasoning

Justification and generalization are closely related to proof in mathematics. In fact, for many mathematicians and in many mathematics curricula, mathematical reasoning is equated with proof. In this book we take the view, together with others (Ball and Bass 2003; Davis and Hersh 1981; Hanna and Jahnke 1996; Kilpatrick et al. 2001; Kline 1980; Krummheuer 1995), that whereas proof is one form of argument and justification, not all arguments and justifications are proofs, and a formal proof is not always an adequate justification or explanation of mathematical ideas. Although formal proof has long been thought to produce infallibility in mathematical knowledge, in fact it does not do so (Davis and Hersh 1981; Ernest 1991; Hanna and Jahnke 1996). Standards of rigour are socially constructed (Ernest 1991; Volmink 1990) and "there has never been a single set of universally accepted criteria for the validity of a mathematical proof" (Hanna and Jahnke 1996, p. 884). For example, most mathematics teachers are convinced by the standard one-page presentation of the proof of Pythagoras' theorem; however, a completely logically rigorous proof would take about eighty pages (De Villiers 1990).

Just as in other disciplines, communities of practice (Wenger 1998) exist in the various domains of mathematics, which review new mathematical proofs in accordance with the current questions, objects of study, ways of thinking, methods, and results of the specific mathematical domain. The nature of the discipline of mathematics, founded and built on fundamental, shared concepts means that there is more agreed upon knowledge in mathematics than in other disciplines, such as psychology or sociology. However, this does not mean that mathematical knowledge is not socially constructed or contested. Proof does not shield us from the uncertainty of our knowledge (Hanna and Jahnke 1996; Kline 1980). At the same time, proof is an important embodiment of mathematical reasoning and needs to be taught as a particular form of reasoning, justification, and generalization within the discipline of mathematics (Hanna and Jahnke 1996).

Creativity and Reasoning

A strong rebuttal to the hegemony of proof in mathematics comes from practising mathematicians, who often work intuitively and creatively, searching for understanding and meaning, rather than rigour and formality. Sternberg and his colleagues distinguish between creative and analytical thinking (Sternberg 1999; Sternberg et al. 1998), arguing that "analytical tasks involve analysing, judging, evaluating, comparing and contrasting, and critiquing; creative tasks involve creating, inventing, discovering, imagining and supposing" (1998, p. 374). Although creative and analytical thinking are often posed as dichotomous, they actually support each other in mathematical problem solving and reasoning, for example imagining would require some form of comparing and supposing usually requires some analysing. Comparing alternative solutions, ideas, and imaginings all require reasoning and justification; creative thinking can support links between previously unconnected ideas; and leaps of imagination are often necessary to see a problem from a different perspective.

Intuition has also been studied as an important part of mathematical problem solving, creating mathematical arguments, and proving mathematical theorems (Fischbein 1987). Intuition might precede more formal arguments, justifications, and proof, and in some instances, might replace it. A mathematician who intuitively feels that something is wrong in a proof, will search to find the mistake, doubting the proof rather than her intuition (Hanna and Jahnke 1996). Crucial to notions of creativity and intuition is a sense that conviction and understanding do not necessarily come from formal, deductive, or analytic proofs. Although these have their place, they are certainly not sufficient to solve mathematical problems and communicate mathematical justifications and generalizations. If practices in the mathematics classroom are to be authentic to the discipline of mathematics (Brown et al. 1989), then a broader range of reasoning should be acknowledged and developed in mathematics classrooms.

Empirical and inductive reasoning play an important part in the reasoning practices of mathematicians and mathematics learners, often complementary to theoretical and deductive reasoning. Simon (1996) argues for a notion of "transformational" reasoning, where dynamic transformations of objects are visualised and which provide the reasoner(s) with a sense of conviction and understanding of how and why something is the case. Transformational reasoning supports and is supported by both inductive and deductive reasoning. Drawing on Toulmin, Krummheuer (1995) argues for substantive arguments, rather than merely analytic ones. Substantive arguments show relationships between the main objects and premises, rather than merely drawing deductive conclusions based on previously proved results or axioms. This distinction is similar to Hanna's characterization of proofs that prove and proofs that explain (Hanna and Jahnke 1996). De Villiers (1990) argues for five key functions for proof – verification, explanation, systematization, discovery, and communication. It is useful to see these as functions of mathematical reasoning as well. Verification establishes that something is the case, i.e. sufficient justification has been produced to confirm that a claim is true. Explanation establishes why something is the case, showing what are the key properties that are necessary for the truth of a claim. Explanatory proofs, or substantive arguments are more satisfying to both mathematicians and mathematics learners (Hanna and Jahnke 1996; Krummheuer 1995). Systematization organizes disparate mathematical concepts that are already established into a coherent mathematical system. As argued above, mathematical reasoning is a key part of mathematical discovery and mathematical reasoning also functions to help communicate our ideas and their warrants to others.

The idea that mathematical reasoning involves creativity, discovery, and communication is central to the work of this book. In Chap. 4, we show how a collaborative conversation among learners supported the development of the mathematical concept of function. Communication was the key in enabling learners to make creative, reasoned conceptual leaps. In Chap. 5, we show how the teacher's practices supported the learners' mathematical reasoning by encouraging them to question and challenge each other and himself. Again, we see reasoned creativity among his learners.

In this section, I have argued that mathematical reasoning is a key element of mathematics and thus is central to learning mathematics in school. I have argued for a broader notion of mathematical reasoning, in which intuition, creativity, imagination, explanation, and communication all play an important role. Fundamental to all forms of mathematical reasoning is the practice of justification and creating adequate arguments in defence of claims. Throughout this section, I have drawn on the notions of mathematical practices, communities of practice, and that mathematics is fundamentally a social practice. In the next section, I explore these ideas further.

Theories of Learning and Mathematical Reasoning

The work in this book is informed by a number of theories of learning, in particular constructivist, socio-cultural, and situated theories. Following Sfard (1998, 2001), I argue that none of the above theories is sufficient on their own to explain the learning and teaching of mathematical reasoning, and in this project I use them in careful combination. Although some scholars argue that since the fundamental mechanisms that generate learning posed by the theories are so different (biological equilibrium for constructivists and social relations for socio-cultural and situated theories), the theories may be incommensurable, my argument is that the different mechanisms operate at different levels and in combination with each other and as long as the differences are acknowledged and specified, we can use these theories together to inform teaching and account for learning in mathematics classrooms (see also Sfard 2001).

Constructivism

Constructivism, in its many varieties, is centrally concerned with how knowledge is constructed and restructured in order to make sense of ever-increasing complexity, both in one's knowledge and in the outside world. Constructivism has had an important influence on theories of mathematics learning and mathematical reasoning (Confrey and Kazak 2006; Hanna and Jahnke 1996), and on the new curriculum in South Africa (Department of Education 2000). However, just as there are many varieties of constructivism, there are many ways in which constructivism can be misconstrued (Moll 2000).

The version of constructivism that informs the work in this book is derived from Piagetian constructivism (Piaget 1964, 1968, 1975), informed by the interpretations of Hatano (1996) and Rowell (1989). Two key principles of this version are first, that what people learn is constrained and afforded by what they know; and second, that there is an integrity to learners' thinking – what learners think, say, and do makes sense to them in relation to what they know. The role of current knowledge is very particular in that current knowledge is not merely built upon (as in behaviourist theories); rather it is restructured and reorganized into richer, more connected, and more powerful knowledge (Hatano 1996). Just as new knowledge is transformed in relation to prior knowledge, so prior knowledge is transformed in relation to new knowledge. From constructivist perspectives a deepening or transforming of thinking involves a deepening or transforming of cognitive structures, either integrating previously separate structures into more general and powerful structures, or differentiating previous structures into more nuanced structures, which allow for more depth of thinking (Hatano 1996). The implication for mathematics classrooms is that teachers need to find out how learners are thinking in order to help them build relationships between current and new knowledge.

The unit of analysis in constructivist theories is the mind of the individual learner. Social interaction is crucial to constructivism in that it supports and constrains individual learning (Hatano 1996; Sfard 2001). However, social interaction is a secondary mechanism, important only as long as it engages the key mechanism for learning and development – equilibration (Piaget 1964; Rowell 1989). Equilibration is a biological process where perturbations to current knowledge structures are compensated for in ways that develop them into more powerful structures. Although the initial perturbation might be created by social interaction, the biological processes must engage for a shift in knowledge to occur. At the same time equilibration on its own is not sufficient to account for learning, because social processes must be taken into account as well.

The concept of cognitive conflict explains the links between biological and social processes in constructivism. Cognitive conflict is where a teacher or peer challenges the position of the learner, illuminating a contradiction in her/his thinking. The theory holds that if the challenge creates a perturbation in the learner, then the learner will equilibrate and develop more powerful knowledge. However, research and experience show that even when learners can see the contradiction, they are often more comfortable maintaining contradictory positions than trying to achieve coherence (Sasman et al. 1998), and might become defensive of their current knowledge (Balacheff 1991; Chazan and Ball 1999). Although constructivism might be able to account for how people do learn, it is less successful in accounting for how they do not learn (Slonimsky, personal communication).

A key part of constructivist research has been work on misconceptions (Confrey 1990; Smith et al. 1993), which, in our experience, has been extremely helpful for teachers. This research shows that learners' errors are often systematic and consistent across time and place, remarkably resistant to instruction, and extremely reasonable when viewed from the perspective of how the learner might be thinking. To account for these "rational errors" (Ben-Zeev 1996, 1998) researchers posit the existence of misconceptions, which are underlying conceptual structures that explain why a learner might produce a particular error or set of errors. Misconceptions make sense when understood in relation to the current conceptual system of the learner, which is usually a more limited version of a mature conceptual system (for this reason, many researchers prefer terminology such as "alternate conceptions"). Misconceptions result from structures that apply appropriately in one domain being over-generalized to another, for example, the idea that you cannot take away a bigger number from a smaller makes sense in the domain of natural numbers, but not in the domain of integers. Thus misconceptions are a normal part of the learning process. Misconceptions have been thought to arise from teaching that emphasizes procedures and individualized instructions (Ben-Zeev 1996; Erlwanger 1975; Schoenfeld 1988). However reports from teachers working conceptually and collaboratively suggest that misconceptions continue to arise in these classrooms

(Ball 1996, 1997; Chazan and Ball 1999; Lampert 2001), which is to be expected, since misconceptions are a normal part of learning.

Since misconceptions form part of the learners' current knowledge, the well established educational truism that teachers need to work with and build on learners' current knowledge suggests that teachers should work with learners' errors and misconceptions as well as their correct ideas. Misconceptions alert us to the fact that "building" on current knowledge also means transforming it; current conceptual structures must change to become more powerful or more applicable to an increased range of situations. At the same time the new structures have their roots in and include earlier limited conceptions (Smith et al. 1993). Learners' misconceptions, when appropriately coordinated with other ideas, can and do provide points of continuity for the restructuring of current knowledge into new knowledge (Carraher 1996; Confrey 1990; Hatano 1996; Smith et al. 1993).

The idea that learners' errors arise from the underlying conceptual structure of the learner and can be an indication of appropriate reasoning and the integrity of the learner's thinking, can be extremely powerful in helping teachers to shift their teaching towards taking learners' thinking seriously (Ball and Cohen 1999; Nesher 1987). Teachers who orient toward learner thinking would want to try to understand the thinking that produces the learners' contributions, including their errors. They would see errors as a normal part of coming to a correct conception. Since misconceptions can also produce correct responses (Nesher 1987), asking learners to explain their thinking when they produce both correct and incorrect contributions is a way to access appropriate or inappropriate underlying mathematical reasoning. In Chap. 8 of this book, I draw on the notion of misconceptions to develop a language of description for learner contributions, which takes learners' errors and partial insights seriously while looking for ways to transform them into more appropriate understanding.

Socio-Cultural Theories

One of the key implications of constructivist theories, which has been popularized in teacher training around the new curriculum in South Africa, is that teachers are "facilitators", which means that although they might support learning through appropriate tasks and questions, they are not directly implicated in it (Department of Education 1997; Hanna and Jahnke 1996). Teachers are often exhorted not to "tell" learners any mathematics (Chazan and Ball 1999), for fear that they might inhibit the learners' own constructions. Socio-cultural theories provide a direct challenge to this view, as they argue that adults, and teachers in particular, as bearers of the culture, must be involved in developing learners' understanding and in so-doing, must leave their mark on what learners learn (Hatano 1996; Sfard 2001). The processes of construction will include much of the teachers' language and ways of seeing that learners appropriate as teachers work with them.

A key difference between socio-cultural and constructivist theories is that socio-cultural theories posit social interaction as the primary mechanism in intellectual development. For Vygotsky (1978), the interpsychological (interaction among people) becomes the intrapsychological (mental functions). Vygotsky has a very strong notion of the mind, which is formed biologically, through the lower mental functions, and socially, through the higher mental functions. He argues that social interaction and broader cultural historical patterns are constitutive of higher order consciousness. The interaction between the social and biological is therefore key, as in Piaget's constructivism; however, the social is primary.

For Vygotsky, social, cultural, and historical knowledge is carried through signs and artefacts and mediated to younger members of the culture by more experienced members. So for Vygotsky, the unit of analysis is always the individual interacting with another person or people, either directly, or through a tool or artefact (for example, a book).

This is formalized in his notion of the zone of proximal development (Vygotsky 1978). In some of Vygotsky's writings, it appears that he conceives of the zone of proximal development as belonging to an individual learner. However, the social nature of his theory suggests that zones of proximal development are created in interaction between learner and teacher or between learner and artefact (Hedegaard 1990; Wertsch 1984). Thus the same teacher/artefact can create different zones of proximal development with different learners, and different teachers/artefacts can create different zones of proximal development for the same learner. Mediation is crucial (Crook 1994; Herrenkohl and Wertsch 1999) in that it creates the conditions of possibility for internalization of the key concepts of the culture.

Many teachers, including those whose work is represented in this book, have found the concepts of the zone of proximal development and mediation extremely appealing, because they posit a central role for the teacher. In Chap. 4, we draw on the socio-cultural theory together with constructivism to show how one learner's development of mathematical reasoning is mediated by conversation with her peers and the teacher, as they use a set of mathematical resources together to solve a problem. We see how the learners' collaborations are intimately connected with and become part of an individual learner's increasingly sophisticated reasoning. In Chap. 5, we draw directly on the zone of proximal development, arguing that the teacher's practices of questioning and listening to the learners' mathematical reasoning form a zone of proximal development for the learners and they begin to listen to and question each other, and him, in similar ways.

It should be noted here that Vygotsky's theory is also a theory of social construction. Interpsychological processes do not become intrapsychological processes without being transformed. Vygotsky states: "adults, through their verbal communication with the child, are able to predetermine the path of the development of generalizations and its final point – a fully formed concept. But the adult cannot pass on to the child his mode of thinking" (1986, p. 120), and Leont'ev writes "the process of internalisation is not transferal of all activity to a pre-existing plane of consciousness; it is the process in which this internal plane is formed." (1981, p. 57 in Cazden 1988, p. 108). So, internal processes, although constituted by external processes, do not mirror them.

There are two strong critiques of Vygsotsky's work, both relating to his notion of internalization. The first is that the processes of internalization are left relatively un-theorised. This is one area where Vygotsky's work remains insufficient[3] and where Piaget's notions of assimilation, accommodation, and equilibration do far more to explain how external ideas are internalized. Related to this is Vygotsky's under-acknowledgment of the role of errors and misconceptions in learning. His work often suggests that learning proceeds relatively smoothly in the zone of proximal development. The second critique is that there is far too much emphasis on internalization in Vygotsky's theory, because of its strong focus on mind (Crook 1994; Daniels 2001; Lave 1993). This critique suggests that a focus on social relations is more useful in understanding learning. Related to this argument is the fact that Vygotsky's theory, while acknowledging necessary asymmetrical relationships between teacher and learner, does not always acknowledge the power differences among learners, particularly in relation to race, gender, and class and how these might affect learning. A more participatory account allows for these to be included.

Situated Theories

Situated theories view learning as participation in communities of practice (Lave 1993; Lave and Wenger 1991; Wenger 1998). To view learning as participation is to say that not only does learning occur through participation, as both constructivist and socio-cultural theories argue, but that learning is defined and identified as increasing participation in a practice. To learn is to participate better. To learn mathematics is to become a better participant in a mathematical community and its practices, using the discursive tools and resources that the community provides (Forman and Ansell 2002; Greeno and MMAP 1998). The unit of analysis in situated theories is the community of practice and it is important to specify the practices of a particular community (Brown et al. 1989). Communities of practice constitute contexts for the learning of their members, and as communities they also learn (Wenger 1998). The mechanism for learning in situated theories is legitimate peripheral participation in communities of practice (Lave and Wenger 1991). Newcomers to the practice participate legitimately, but on the periphery, at first. As they gain experience, their participation shifts towards full participation, which is their learning. As newcomers become oldtimers, so the community itself learns and shifts, creating both personal and communal growth. The processes of negotiation between newcomers and oldtimers can create tensions and conflict, as newcomers

---

[3] I thank Steve Lerman for pointing out that this is probably because Vygotsky died a short time after formulating the notion of zpd and internalization.

stake their positions in the community. Thus power relations, which are not taken into account by constructivist and socio-cultural theories, become the key to learning in situated theories. Lave and Wenger argue that their theory of social practice "emphasizes the inherently socially negotiated character of meaning and the interested, concerned character of the thought and action of persons-in-activity" (Lave and Wenger 1991, p. 50).

Similar to socio-cultural theories, situated theories view the role of the social interaction as constitutive of learning; they are social theories of learning (Wenger 1998). Different from socio-cultural theories, they view learning as only a social phenomenon; the definition of learning as participation rather than *constituted by* participation suggests learning is social and not mental. Learning is the creation of identities in communities of practice (Wenger 1998). Situated perspectives argue that knowledge cannot be seen as the "possession" of an individual, but rather is distributed among people and resources (Greeno et al. 1996; Sfard 1998).

In situated perspectives, a concern with thinking is transformed into a concern with participation, with how learners use mathematical tools and discourse to reason and justify their reasoning (Sfard 2001). Making connections and generalizing ideas are important in situated perspectives, however the connections and generalizations are ideas that are in the conversation, rather than structures in the head. Greeno and MMAP (1998) suggest that situated analyses broaden the notion of conceptual structures into one of attunement to affordances and constraints. Affordances and constraints are located in interactional situations, in the classroom, the domain of mathematics, and the lives of the learners beyond school. From this perspective, learners who do well in mathematics do so because they align and identify with the requirements and expectations of the classroom, both mathematical and social. A learner may struggle to learn, not because she has not developed appropriate conceptual structures, but because she is responding to a different task than the one set by the teacher, or does not want to be seen to be too intelligent or not intelligent enough in front of her peers, or has decided not to engage with mathematics because it does not seem to be important in the lives of people who are important to her. Such attunements are patterned regularities, which may be just as important in accounting for learning as are conceptual structures.

The de-emphasis of conceptual structures as products of learning makes situated theories somewhat difficult for teachers to own and work with, epecially teachers, like the teachers in this book, for whom conceptual understanding and mathematical reasoning are important. However, there is one key notion in situated theories – communities of practice – which attract teachers, and which seem to be compatible with the key elements of socio-cultural and constructivist theories. Communities of practice present images of how communication can take place in classrooms, the roles of resources, of different learners, and of the teacher. Although classroom communities must be somewhat different from communities outside of classrooms (Lave 1993, 1996; Lerman 1998), classroom communities, where genuine mathematical communication and the development of mathematical understanding and identities take place, can be established (Boaler 2004; Boaler and Greeno 2000; Boaler and Humphreys 2005; Lampert 2001; Staples 2004).

This view infuses the work in Chaps. 4–6 of this book, where we show how communities of practice were created in three of the classrooms. It also informs Chaps. 8 and 9, which show how learner contributions and teacher responses co-produce each other, and Chaps. 10 and 11, where dilemmas of teaching and resistance to teacher change are viewed as profoundly situated and developed in and through communities of practice.

Teaching Mathematical Reasoning

The theories discussed above are primarily theories of learning. It is often thought that theories of learning have direct application to classrooms and suggest particular pedagogical approaches. However, this is not the case; rather theories of learning suggest general pedagogical principles and implications for pedagogy; they do not directly lead to particular pedagogical approaches. Moreover, pedagogical principles do not derive from theories of learning in a one-to-one correspondence. Different theories might suggest very similar approaches, which are distinguished at the level of explanation rather than at the level of practice. It might be tempting to conclude that since constructivist theories focus on the individual, they suggest individual approaches to teaching and learning, while socio-cultural and situated perspectives suggest group work. However, all three theories suggest that group work is a useful pedagogical approach and none would advocate that learners do no work on their own. From all three perspectives, encouraging learners to talk through their ideas with each other is an important process, as is encouraging learners to write down different versions of their thinking, for themselves and others. Constructivist perspectives suggest that when learners are pushed by others to articulate their thinking, they are likely to clarify their thinking, both for others and for themselves (Barnes and Todd 1977; Glachan and Light 1982; Mercer 1995; Vygotsky 1986). In the process of clarifying their thinking, learners might develop more complex concepts, through differentiation, integration, and restructuring (Hatano 1996). Situated perspectives suggest that as learners consider, question, and add to each other's thinking, important mathematical ideas and connections can be co-produced. For constructivist perspectives the group is a social influence on the individual; for sociocultural and situated perspectives the group is the important unit, which produces mathematical ideas within or beyond the individual learner. One, or both of these purposes for group work might be operating in a classroom at any particular time.

In this section, I delineate pedagogical implications for teaching mathematical reasoning, drawing on the arguments in the previous sections of this chapter. The key in teaching mathematical reasoning, as in teaching any other aspect of mathematical proficiency, are the kinds of tasks that learners engage in, the ways in which they engage with these tasks, and the kinds of interactions around the tasks among the learners and the teacher. However, as noted by Ball and Bass, "simply posing open-ended mathematical problems that require mathematical reasoning is not sufficient to help students learn to reason mathematically. Neither is merely asking students to explain their thinking" (2003, p. 42).

Tasks for Mathematical Reasoning

A number of frameworks have been developed to describe the complexity of tasks (Biggs and Collis 1982; Shavelson et al. 2002; Stein et al. 1996, 2000). In this project we drew mainly on the work of Stein et al. (1996, 2000). Although we acknowledge limitations with this framework (Sanni 2008a), it was very useful as a starting point for teachers wanting to select tasks to support learners' mathematical reasoning. This framework is discussed in detail in Chap. 3, where we use it to analyse how learners responded to tasks intended to develop mathematical reasoning. For the purposes of this chapter it is sufficient to note that Stein et al. identify task features which support higher cognitive demands on learners, including reasoning and sense making. These features are "the existence of multiple-solution strategies, the extent to which the task lends itself to multiple representations, and the extent to which the task demands explanations and/or justifications from the students" (Stein et al. 1996, p. 461). Stein and others (Ball 1993; Boaler and Humphreys 2005; Chazan 2000; Lampert 2001) show that tasks that support multiple voices, disagreements, and challenges also support mathematical reasoning, when used appropriately. Douek (2002) argues for specific kinds of complexity in tasks to support the development of mathematical arguments, including the complexity of integrating a number of different arguments into a coherent whole, the complexities involved in moving from dynamic to static representations, and the complexity of the contexts in which tasks are set. Garuti and Boero (2002) describe teaching experiments where the arguments of famous scholars (Galileo, Plato) are presented to students as examples of forms of argument, and students are asked to write similar arguments for mathematical problems. Considering the arguments of others, including those of one's own peers, can be a powerful source of developing a learner's own reasoning and arguments.

Choosing appropriate tasks is necessary but not sufficient to support a learner to develop reasoning. Stein et al. (1996) show that, with support, the teachers in their project chose tasks that made higher order cognitive demands on learners. However, as the tasks were implemented in the classrooms, the level of demand declined. In South Africa, Modau and Brodie (2008) show how a teacher teaching the new curriculum in Grade 10, supported by a new curriculum textbook, chose tasks that required reasoning. However, at implementation, he was not able to maintain the level of the tasks, but through his questioning and patterns of interaction, lowered the task demands and thus did not support reasoning (Jina and Brodie 2008). Sanni has shown that in six Nigerian classrooms the level of most of the tasks also declined. However, when he worked as a support for one of the teachers, the level of the tasks remained high and the learners' reasoning improved (Sanni 2008b).

Taken together these studies suggest that substantial work with teachers is required to support them to interact with their learners on tasks to support the learners' mathematical reasoning.

Classroom Interaction

Since communication is fundamental to reasoning (Ball and Bass 2003; Douek 2005; Krummheuer 1995), it makes sense that learners should discuss their reasoning with others. This is supported by all the learning theories discussed previously and by curriculum reforms in South Africa and internationally. As learners attempt to create reasoned arguments for their ideas, they help themselves and each other to clarify their thinking and they are able to create some of the practices that mathematicians engage in as they produce arguments and justifications. However communication can be structured in a variety of ways, leading to very different kinds of support for mathematical reasoning. Many South African teachers talk about the "question and answer method" as if this guarantees learner participation. However, it is well known in the research literature that if the questions are narrow and do not challenge learners' thinking, then the resulting interaction is stilted and does not support reasoning (Bauersfeld 1988; Edwards and Mercer 1987; Mehan 1979; Nystrand and Gamoran 1991). On the other hand, putting learners into groups and leaving them to work without mediation from the teacher does not necessarily provide enough support for developing their reasoning (Brodie and Pournara 2005). Even whole class discussions are often not successful, because working out exactly how to respond to the learners' developing ideas and reasoning is a difficult task for teachers (Heaton 2000).

The work on classroom interaction in Chaps. 8–10 of this book draws on work done many years ago by Mehan (1979) and Sinclair and Coulthard (1975) on the pervasive exchange structure of classroom discourse, the Initiation-Response-Feedback/Evaluation (IRF/E) exchange structure. The teacher makes an *initiation* move, a learner *responds*, the teacher provides *feedback* or *evaluates* the learner response and then moves on to a new *initiation*. Often, the feedback/evaluation and subsequent initiation moves are combined into one turn, and sometimes the feedback/evaluation is absent or implicit. This gives rise to an extended sequence of initiation-response pairs, where the repeated initiation works to achieve the response the teacher is looking for. When this response is achieved, the teacher positively evaluates the response and the extended sequence ends.

Neither Sinclair and Coulthard nor Mehan evaluated the consequences of the IRE structure. Other researchers (Edwards and Mercer 1987; Nystrand et al. 1997; Wells 1999) have argued that it may have both positive and negative consequences for learning. Although this structure requires a learner contribution every other turn (the response move), and therefore apparently gives the learners time to talk, much research has shown that because teachers tend to ask questions to which they already know the answers (Edwards and Mercer 1987) and to "funnel" learners' responses toward the answers that they want (Bauersfeld 1988), space for genuine learner contributions is limited. At the same time, it is very difficult for teachers to move away from this structure (Wells 1999) and so, in trying to understand a range of pedagogies, it is important to try to understand the benefits that it affords. Whether the IRE has positive or negative consequences for learning will most likely depend on the nature of the elicitation and evaluation moves, which in turn influence the depth and extent of the learners' responses. In Chap. 9, I develop a language to describe teacher responsiveness in the fused elicitation/evaluation moves, which distinguishes a number of teacher moves and their consequences for learner contributions.

One aspect of classroom interaction that has been identified as important to supporting useful interaction, is the development of ground rules (Edwards and Mercer 1987) or classroom norms (McClain and Cobb 2001), which are different from those in traditional classrooms. Norms that support reasoning would includethe following: learners are called on to justify all their reasoning, not only mistaken reasoning as might happen in traditional classrooms; learners are expected to listen to and build on each others' ideas and challenge them where necessary; learners can and should challenge the teacher, and the teacher should justify her/his mathematical thinking. This raises the important issue of authority in mathematics classrooms. Traditionally, mathematics learners are expected to accept the authority of the teacher or the textbook, rather than the authority of mathematical justifications. These two kinds of authority are very different (Brousseau and Gibel 2005), and are implicated in the learners developing a productive disposition towards mathematics (Kilpatrick et al. 2001) and hence achieving overall mathematical proficiency. Productive disposition is a belief that mathematics can and does make sense, and that every learner can make sense of it. This requires an understanding that the "rules" are not arbitrarily decided on by powerful individuals, but that they make sense in terms of a system of knowledge, which can be understood by everyone, with sufficient effort.

Ball and Bass (2003) argue that mathematical justification is grounded in a body of public mathematics knowledge, where this knowledge can be that of a group of mathematicians, or an elementary classroom community. This public knowledge ensures that individual sense making becomes accountable to a broader community; because an idea making sense to an individual is not the same as an idea being based on shared reasoning in communities of mathematicians over time. So taking the individual learner's reasoning seriously means attempting to connect it to accepted mathematical reasoning. They argue, "reasoning, as we use it, comprises a set of practices and norms that are collective, not merely individual or idiosyncratic, and rooted in the discipline" (p. 29). It follows from their position, although they do not argue it, that even the classroom mathematics community cannot be the final arbiter on the acceptance of a mathematical argument, because this community is accountable to broader communities of mathematical practice and to the discipline of mathematics. The idea of accountability to the discipline is one that attracted all of us in this project, and subsequent chapters will show how we worked with this notion and how the learners received it.

The Challenges of Teaching Mathematical Reasoning

In this chapter, I have outlined the theoretical positions on mathematics, learning and teaching, which inform this book and have suggested some of the many challenges that teachers face when trying to make reasoning a central part of their practice. I have also suggested some challenges that researchers and teacher educators face when looking for ways to talk about the teaching and learning of mathematical reasoning. In the rest of this book, we take up some of these challenges:

- How do learners respond to tasks chosen to elicit their mathematical reasoning?
- How can teachers interact with learners around tasks to engage their mathematical reasoning?
- How can teachers teach to develop mathematical proficiency?
- How does collaborative conversation among learners and teacher promote mathematical reasoning?
- What kinds of teaching practices, questions, and moves help to encourage and sustain the learners' mathematical reasoning?
- How can we describe the learners' contributions and teachers' responses to these in ways that can help us talk about them in more specific ways?
- What kinds of dilemmas do teachers experience as they teach mathematical reasoning?
- What can teachers do in response to the resistance to new ways of teaching?

We approach these questions from two directions. In part two of this book, we present five studies conducted by teachers in their classrooms, each of which addresses one or two of these questions. The range of contexts in which the teachers work and their approaches to the topic of mathematical reasoning suggest a number of possibilities to other teachers, teacher educators, and researchers wanting to undertake case-studies of teachers teaching mathematical reasoning. In part three of the book, I look across the teachers' practices, in a multiple-case study. I suggest a language of description for learner contributions and teacher moves, and I use this language to illuminate ways in which teachers, teacher educators, and researchers can gain deeper insight into how to respond to the learners' mathematical thinking. I also use this language to illuminate some of the dilemmas that teachers experience when engaging learners' mathematical reasoning and to talk about the resistance that they may experience.

It is well established that meaningful change in teaching and learning takes time. In this book we illuminate both successes and challenges in teaching for mathematical reasoning, among ordinary teachers, to give substance to why such teaching takes time to achieve. We do not claim to have succeeded in producing the ideal teaching and learning situations that reformers might hope for, and we are not sure whether such perfection is possible. We do claim to have learned much about what we have achieved and how we might move forward. We hope that our work in developing research and practice together will provide ideas and possibilities for many other teachers, teacher-educators, and researchers to begin their own explorations in teaching mathematical reasoning.

Chapter 2
Contexts, Resources, and Reform

In the previous chapter, I outlined possibilities for teaching mathematical reasoning that involves learners communicating their thinking to their teacher and their peers and teachers taking learners' mathematical reasoning seriously to develop and transform it. This is in line with the visions of reform mathematics in a number of countries. However, international evidence suggests that very few teachers embrace reforms and those who do, experience significant challenges in their teaching. The challenges that I outlined at the end of the previous chapter are daunting in any context even in the most well resourced contexts. However, in South Africa and many other countries, resources in most schools are severely limited, adding to teachers' difficulties in enacting reforms.

At the same time, resources are not the only influence on reform teaching and the ways in which they exert an influence are not always obvious. In a review of recent international and South African studies, Fleisch (2007) shows that the studies are inconclusive on the effects of resources such as teacher qualifications, class size, and learning materials on learner achievement. This suggests that there are ways in which resources that do matter are most likely mediated by other variables. In this chapter, I discuss some responses to reform pedagogy across a range of contexts and discuss ways in which contextual constraints and resources may or may not be implicated in enactments of reform teaching. This discussion, together with a more general discussion of the resources available for teaching and learning in South Africa, serves to situate the description of the different contexts of the teachers in this study and the resources available to them as they worked to teach mathematical reasoning in their classrooms.

Responses to Reforms

A strong impetus for reform curricula in many countries is the need to redress inequalities in mathematics education. Internationally, success in mathematics is distributed according to race and socio-economic status (Association for Mathematics Education of South Africa 2000; Department of Education 2001;

K. Brodie, *Teaching Mathematical Reasoning in Secondary School Classrooms*,
DOI 10.1007/978-0-387-09742-8_2, © Springer Science+Business Media, LLC 2010

Moses and Cobb 2001; Secada 1992). While many of the reasons for this maldistribution originate outside of the classroom, there are arguments that classroom practices can begin to work towards equity (Boaler 2002). Allowing different ways of knowing mathematics to be available in the classroom may afford success for a wider range of learners (Boaler 2002; Boaler and Greeno 2000). Allowing learners to express their ideas in the classroom can lend to diverse ways of thinking, and help to teach learners that everyone's thinking can contribute to the development of mathematical knowledge (Lampert 2001).

There has been much debate as to whether current mathematics reforms can be a mechanism for ensuring more equitable participation and achievement in mathematics (see Brodie 2006, for a summary of these debates). Empirical evidence in well-resourced countries is beginning to show that reforms do mitigate achievement gaps between marginalised and other learners and also enable learners to develop more motivated and positive identities as mathematics learners (Boaler 1997; Boaler and Greeno 2000; Hayes et al. 2006; Kitchen et al. 2007; Schoenfeld 2002). However, the evidence also shows that implementation of reform curricula is not widespread and in fact it is likely that implementation of reforms is inequitably distributed (Kitchen et al. 2007), so that poorer learners are less-likely to experience reform curricula and pedagogy. Particularly in African contexts, issues of resources, including big classes and few materials, teacher confidence and knowledge, and support for teachers, can be major barriers to developing new ways of teaching (Tabulawa 1998; Tatto 1999). If reforms are successful in promoting equity and if they are not taken up in less-resourced contexts, then the existing division between rich and poor are likely to be exacerbated.

There is also growing evidence that teaching in reform-oriented ways is an extremely challenging task for teachers (Sherin 2002; Nathan and Knuth 2003) and that successful reform teachers are rare, even in well resourced schools in the United States where the reforms have been in place for 10 years longer than in South Africa. Among the 18 teachers in their study, Fraivillig et al. (1999) considered only six teachers to be "skillful" in eliciting and supporting learner thinking, while only one was successful in eliciting, supporting, and extending learner thinking. Hufferd-Ackles et al. (2004) described the development of reform practices through four levels. Of the four teachers they worked with, only one teacher's trajectory took her and her learners through all four levels. These studies were conducted in well-resourced classrooms and so suggest that while resources may be important, they are not the only challenge for teachers in working with learners' reasoning.

Studies of and by teachers who are successful in developing discussion and collaboration around learners' reasoning, identify a number of challenges in such work. These include: supporting learners to make contributions that are productive of further reasoning (Heaton 2000; Staples 2004); respecting and valuing all learners' thinking while working with the diversity of their mathematical ideas (Lampert 2001); respecting the integrity of learners' errors while trying to transform them and teach the appropriate mathematics (Chazan and Ball 1999); seeing beyond one's own long-held and taken-for-granted mathematical assumptions in order to hear and work with learners' reasoning (Chazan 2000; Heaton 2000); maintaining a "common ground", which enables all learners to follow the conversation and its mathematical purpose and to contribute appropriately (Staples 2004); and generating mathematical practices such as making connections, generalizing, and justifying (Boaler and Humphreys 2005). The above research shows that the pedagogical demands of mathematical conversations can be daunting and that we need to understand more about the practices involved in generating and sustaining these conversations (Brodie 2007b).

Research on the new curriculum in South Africa has shown that teachers who are enthusiastic about and express support for the new curriculum struggle to enact many of the ideas in their classrooms. These studies show that many classrooms remain teacher-centred, and teachers engage with learners' ideas in superficial ways, if they do so at all (Chisholm et al. 2000; Taylor and Vinjevold 1999). Other studies show some hybrid practices beginning to develop. Jansen (1999) found that while most teachers were not implementing the new curriculum, or were doing so very superficially, some of the more experienced and confident teachers were able to move between old and new practices and negotiate for themselves and with their learners what it means to implement the new curriculum. Brodie et al. (2002) found that many teachers set up tasks and group work situations where learners engaged with the tasks. However, when learners expressed their thinking, teachers struggled to support and engage with their reasoning to take them further and to develop them mathematically (Brodie 1999; Brodie et al. 2002). This finding was confirmed in more recent studies, where teachers selected tasks that could elicit mathematical reasoning, but did not engage learners' reasoning in classroom interaction (Jina and Brodie 2008; Modau and Brodie 2008; Stein et al. 1996; Stein et al. 2000).

There are a number of possible explanations for South African teachers' difficulties with the new curriculum. One claim, prominent at the moment, is that teachers do not know enough conceptual mathematics to teach in ways required by the new curriculum (Taylor and Vinjevold 1999). Other explanations are that teacher development around the new curriculum has been inadequate and that appropriate curriculum materials are not available (Chisholm et al. 2000; Taylor and Vinjevold 1999). A third possibility is that teachers are able to implement some aspects of the new curriculum, for example, higher level tasks, relatively easily, whereas other aspects, in particular, interaction with learners, are particularly difficult (Brodie et al. 2007). Slonimsky and Brodie (2006) argue that new curriculum practices require that teachers coordinate a complex set of contextual and knowledge constraints and that such coordination takes a long time to develop. All the above explanations acknowledge that resources are only part of the problem.

One of the aims of this project was to explore possibilities for developing learners' mathematical reasoning in a range of South African contexts and thereby begin to develop a deeper understanding of how resources are implicated in such practices. The next section presents some background on educational resources and achievement in South Africa and the following section describes the differently resourced contexts in which the five teachers in the project worked. In the discussion of resources, I include the material resources of the schools, the human resources, in particular teacher and learner knowledge, and finally the resources that the teachers chose to work with in the study – the tasks that they developed to engage their learners in reasoning mathematically.

The South African Context

As with all aspects of life in South Africa, the education system is characterized by large disparities between rich and poor, and most of our schools and learners are of very low socio-economic status. Most teachers in South Africa teach big classes in very poorly resourced schools. The latest national data shows that of the 25,145 operational schools in South Africa, 11.5% of schools did not have water, 16% had no electricity, 5.2% did not have ablution facilities, 80% did not have libraries and 67% did not have computers for teaching and learning (Department of Education 2007). The average learner–teacher ratio in schools was 32:1 and the average learner-classroom ratio was 38:1 (Department of Education 2008). These averages hide wide disparities between provinces and schools. A research study conducted in Gauteng and Limpopo (the richest and poorest provinces in South Africa respectively), observed averages of 35 students per class in secondary schools, with the three rural secondary schools averaging 60 learners per class and with some classes having as many as 120 learners (Adler and Reed 2002). There were limited resources such as overhead projectors and those resources that did exist, for example chalkboards, were often in poor condition.

While outrageous in any terms, this lack of resources is particularly significant for teachers attempting to work with their learners' mathematical reasoning. It is difficult to attend to learners' ideas when there are 50 or more learners in the class and few material resources. Moreover, many learners, having experienced poorly resourced education, often have weak mathematical knowledge (Fleisch 2007), and may be reluctant to participate in lessons. When they do participate, they may express barely coherent, or very problematic ideas, and teachers may not be able to engage with these ideas (Brodie 2000). The fact that most teaching and learning takes place in English, which is not the main language of most teachers and learners, also makes participation more difficult for learners and development of learner thinking more difficult for teachers.

Learners' weak mathematical knowledge is apparent in the annual results of the school leaving examinations[1], which are taken by approximately 500,000 Grade 12 learners each year, of which about 300,000 take the mathematics examinations. In 2005, 55% of about 303,000 learners passed mathematics and in 2006, 52% of about 317,500 learners passed mathematics. In 2005, about 44,000 learners took the examination at the Higher Grade level, which is required for entry into scientific fields at university, and 59% passed while in 2006, about 47,000 took the examination at this level and 53% passed (Department of Education 2008). The inequities of the system become apparent when we see that of 40,000 Higher Grade candidates in 2001, only 20,000 were "African",[2] and of these about 3,000 passed. Thus,

---

[1] South Africa has school learners examinations at the end of Grade 12. These examinations are high stakes and determine students' eligibility for further study and job opportunities.

[2] The apartheid system of racial classification was four-tiered and funding for education was directly linked to these tiers. Although many white South Africans refer to themselves as African, in this context "African" refers to black South Africans who are not "coloured" or of Indian descent, and whose schools and colleges received least funding under apartheid. Black Africans make up about 80% of the South African population.

while about 85% of white Higher Grade candidates passed, only 15% of black African candidates did. Figures for 2004 are that 7,236 African learners passed out of a total of 40,000 candidates[3] and of these, 2,406 African learners passed with a "C" grade or higher, which is 0.5% of the total number who wrote mathematics and 6% of those who wrote Higher Grade mathematics (Centre for Development and Enterprise 2007).

Research in the lower grades shows that learners begin to struggle with mathematics as early as Grade 3. Contrary to other developing countries, South African learners are almost all in school. Fleisch et al. (2008) show that in 2007, more than 95% of children of compulsory school age attended an educational institution and that this reflects an improvement since 2001 in each age cohort between 7 and 15 years of age. However, Motala and Dieltiens (2008) raise questions as to what these learners actually learn in school, suggesting that about 60% are disengaged and disaffected and learn very little.

Reviewing the research in mathematics, Taylor et al. (2003) conclude that "studies conducted in South Africa from 1998 to 2002 suggest that learners' scores are far below what is expected at all levels of the schooling system, both in relation to other countries (including other developing countries) and in relation to the expectation of the South African curriculum." Many Grade 3 learners struggle with basic skills such as adding and subtracting two-digit numbers that require "carrying" or "borrowing". South Africa also performs poorly in international comparison studies. In the third TIMSS study, South Africa came last out of 41 countries at all three grade levels tested, doing significantly worse than countries with similar GDPs, for example Latvia and Lithuania (Howie and Hughes 1998). In the 2003 TIMSS study, South African Grade 8 learners again came last out of 46 countries, and more significantly did worse than countries with lower Human Development Indices, including Ghana et al. (Reddy 2006). In the Monitoring Learner Assessment Study (MLA), which compared Grade 4 learners across 12 African countries, South Africa had a mean score of 30% for numeracy, which was the lowest of all 12 countries (Taylor et al. 2003). These average scores hide the large disparities between black, low socio-economic status learners and wealthier, white learners, but they serve to show the extent of the "crisis" in mathematics learning in South Africa, which is quantitatively different from many other countries. While we acknowledge critiques and limitations of such comparative studies (for example, Keitel 2000; Reddy 2006), they do serve as an indication of some of the challenges that our education system faces and the difficult conditions under which many teachers work.

The "crisis" also extends to the availability and quality of mathematics teachers in South Africa. Many South African teachers, because they were under-served by apartheid education, have relatively weak knowledge of mathematics and how best to teach it (Taylor and Vinjevold 1999). This situation does not look set to change in the near future. Given the limited numbers of students who graduate from school with strong mathematical knowledge, the pool for potentially well-qualified

---

[3] The number of African candidates who wrote is not available.

teachers is small. Students who do well in mathematics and science usually have a range of more attractive career options in other fields. Knowledgeable teachers of mathematics are often recruited by industry with far better salaries and working conditions. There is a lack of detailed data about mathematics teachers in South Africa (Centre for Development and Enterprise 2007) but the following give only a part of the picture over the past 10 years. In 1997, only 50% of teachers of mathematics had specialized in mathematics in their training (Department of Education 2001); in 2004, a survey of 1,766 secondary schools (out of a total of about 5,600) showed that there were 1,734 qualified mathematics teachers at these schools and of those only 1,362 were actually teaching the subject (Centre for Development and Enterprise 2007); and in 2006, 16 universities graduated a total of about 550 mathematics teachers (Centre for Development and Enterprise 2007).

I have taken some time to review these statistics because they provide a background for understanding the debates about the new curriculum in South Africa, and the contexts of the schools in this study. It is imperative to provide access to mathematics for large numbers of low socio-economic status learners. The government's response has been the development of policies that encourage the transformation of the curriculum and pedagogy in South African schools. The curriculum was developed in consultation with local and international experts and draws on the international research that suggests that reform pedagogy can reduce inequality in mathematics achievement. However, the international experience of teacher difficulties with reform curricula, in addition to the particular challenges of the South African situation suggest that we cannot assume that the new curriculum will reduce inequality – it may even increase it. This was a concern for all of the teachers in this project, and so we wanted to examine how teachers worked with learners' mathematical reasoning in a range of South African classrooms.

Five Schools: Contexts and Resources

Race and Socio-Economic Status

Fifteen years after the end of apartheid, although there have been some shifts, schools still largely reflect historical divisions of race and class. Table 2.1 gives a description of each of the five teachers' schools in terms of race and socio-economic status[4].

---

[4] Historically race and class have been closely related in South Africa. Although this is changing for some small sections of the population, to a large extent this trend still exists and is largely reflected in this sample. In order to determine socio-economic status, we used the location of the school, the school fees, which public schools in South Africa are allowed to charge, and the teachers' knowledge of the typical occupations of the parents.

Table 2.1 Demographics of schools

Teacher	Learners' race[a]	Learners' socio-economic status	School fees (per year)	No. of teachers/ no. of learners
Mr. Nkomo	Black	Working class	R200	1,650/42 (39:1)
Mr. Mogale	Black	Working class	R200	1,700/46 (37:1)
Mr. Daniels	All races	Middle and lower-middle class	R4,000	1,600/60 (27:1)
Mr. Peters	Black and "coloured"	Working class	R400	1,250/36 (35:1)
Ms. King	White, with a few learners of other races	Middle and upper-middle class	R40,000 (private)	850/65 (13:1)

[a]We use apartheid terminology to describe race, which is standard practice in South Africa to indicate shifts or lack thereof in historical racial divisions

Under apartheid Mr. Nkomo's and Mr. Mogale's schools served only black learners, Mr. Peters' school served only "coloured" learners, and Mr. Daniels' and Ms. King's schools served only white learners. Ms. King's school is a private, boys-only school, with some boys living at the school, and all the others are public, co-educational non-residential schools. The racial profile of the teachers matched those of the learners. Since schools began to integrate in the early 1990s, Mr. Peters' school has black and "coloured" learners; Mr. Daniels' school is racially diverse with learners from all four racial "groups"; and Ms. King's school has a few black, "coloured", and Indian learners, but is still predominantly white. Teacher diversity across the schools has occurred much more slowly. All the teachers in Mr. Nkomo's and Mr. Mogale's schools are black, most of the teachers in Mr. Peters' school are "coloured", with some black teachers, almost all of the teachers in Ms. King's school are white, with a few teachers of other races. Mr. Daniels' school has made the most progress in integrating teachers across race. Although many teachers are still white, there are a number of "coloured", Indian and black teachers at the school. Mr. Daniels' himself is "coloured", and moved to this school from a "coloured" school about 4 years ago.

School Resources

All of the five schools are known in their areas as good schools. They all regularly achieved pass rates of 65% and above in the Grade 12 examinations, well above average nationally, and a little above average for Gauteng province in which they are located (Motala and Perry 2002). They are all functional most of the time, in contrast to many other schools in South Africa (Christie and Potterton 1997; Taylor et al. 2003; Taylor and Vinjevold 1999). This means that school starts on time, most learners are present, absentees are noted, learners move between classes relatively quickly, teachers come to class and teach, learners return to classes after breaks, there are regular teacher meetings, and there are administrative staff and administrative computer systems. According to the teachers, the principals are supportive of efforts to improve teaching and learning in their schools. All the principals supported the teachers studying further and all were eager for this research to take place in their schools. So, while the five schools represent some diversity, they do not capture the full diversity of South African schools. The study is limited to an urban area, in the wealthiest province in South Africa, and well-functioning schools. However, as the first study of teaching mathematical reasoning in South African high schools, it was important and appropriate to limit our study in this way.

All of the five schools have fences or walls and control access to the schools[5]. Ms. King's school is located on a beautiful, large, peaceful property, situated next to a busy business district. There are dormitories for the learners who live on campus, houses for the teachers who live on campus, a church, plenty of sports fields and a dam with guinea fowl. It is easy to forget that you are in the middle of a big city while at the school. Mr. Daniels' school is located in a residential suburb, on a hill with a beautiful view of neighbouring suburbs. It has a number of sports fields and has a "green" feel to it. Mr. Nkomo's, Mr. Mogale's, and Mr. Peters' schools are located in residential areas which are poorer and less "green". There is little open space in Mr. Peters' school and some dusty fields in the other two schools. Mr. Peters' school is an area that is well known for gang activity and violence. Sometimes learners come to school with knives and there have been some incidents with guns. During the time of the study, a teacher was robbed at gunpoint of his laptop and cellphone in the school grounds, an incident that caused considerable disruption to teaching and learning for about a week. Learners also are subject to attack when they leave the school, especially many black learners who come from other areas and have to walk through "coloured" gang territory in order to get home. Table 2.2 indicates other resources available at each of the teacher's schools.

Table 2.2 Resources available at the schools

Teacher	Staffroom	Library	Computer room	Photocopying
Mr. Nkomo	Yes	Old books: mainly: textbooks:	Non functional	Yes
Mr. Mogale	Yes			Yes
Mr. Daniels	Tea and coffee	Well equipped	Yes	Yes
Mr. Peters	Yes	No	Non functional	Yes
Ms. King	Tea, coffee, computers	Well equipped	Yes	Yes

---

[5] Christie and Potterton (1997) argue that this contributes to the functionality of schools.

Classroom Resources

Each teacher chose one Grade 10 or 11 class in which we would conduct the research. Table 2.3 gives an overview of the classes that comprised the research sample.

The above table shows the disparities in class size among the teachers, in relation to the socio-economic status of learners at the schools. The schools of lower socio-economic status tend to have larger classes. The only class that does not fit the trend is Mr. Nkomo's, where mathematics classes are smaller in Grade 11 and 12. There is a difference of almost 20 learners between Mr. Peters' and Ms. King's classes. The levels of the classes relate both to tracking practices at the schools and the level of examination that learners are being prepared for. Some schools teach standard and higher grade learners in the same class (Mr. Nkomo's school in this study) while some differentiate them (Mr Mogale's and Mr. Peters' schools). Some schools track even further beyond this (Mr. Mogale's and Ms. King's schools).

Ms. King's classroom is part of a newly built wing of the school, is carpeted and has air-conditioning. There is a big table and chair for each learner, which can be arranged for work in groups. There are whiteboards and pens, a teacher's desk with a computer, cupboards and tables for storing paper and worksheets, notice boards filled with math posters, an overhead projector and screen, and a television set which can be used for presentations from a computer. Each learner has a textbook, which is purchased by the learner. Ms. King has a range of texts and resources, including international texts, from which she and her colleagues develop and share worksheets[6]. Learners have access to a computer lab and so they are given projects to do, either using the mathematical software or using the internet as a research tool, for example a project on the history of mathematics. Mr. Daniels' classroom has small tables and chairs, which are arranged in groups of four. The classroom is in good repair, and there are notice boards with a few math posters that Mr. Daniels has obtained. There is a teacher's desk and one cupboard that overflows with supplies, worksheets, learners' work, and other documents (there is no other storage space).

Table 2.3 Description of research classes

Teacher	Grade	Class size	Tracking/level of class
Mr. Nkomo	11	28	Untracked: mixed standard and higher grade[a]
Mr. Mogale	11	43	Tracked: higher grade (top class)
Mr. Daniels	11	35	Tracked: standard grade
Mr. Peters	10	45	Untracked
Ms. King	10	27	Tracked: second highest class of seven

[a]At the time of the study, the Grade 12 mathematics examination could be taken at two levels, standard and higher grade. Success on the higher grade (or exceptionally good marks on the standard grade) granted access to scientific fields at university

---

[6] Two mathematics teachers at this school developed a very strong curriculum of investigations in the 1980's and Ms. King and her colleagues still use some of this (McLachlan & Ryan, 1994).

There is a chalkboard and chalk and an overhead projector and screen, although electricity is not always available and so it does not always work. The school issues a textbook to each learner, although Mr. Daniels and his colleagues work predominantly from worksheets, which they develop and share.

Mr. Nkomo and Mr. Mogale's classrooms are similar. Tables are shared between two learners, with enough space for both. These are put together in pairs to form groups of four learners. There is a chalkboard and chalk, but no teacher's desk and no overhead projector and screen. Mr. Nkomo's classroom has a cupboard that stores cleaning supplies, but he has an office nearby where he keeps texts, worksheets, and learners' work. Mr. Mogale keeps his work in the staffroom. Mr. Nkomo's classroom is in good repair, although there is graffiti on the cupboard and notice boards. He occasionally puts posters and learners' work on the notice boards but has to be careful because they often disappear. Electricity is regularly available. In Mr. Mogale's classroom, some windows and the door are broken. There is no regular supply of electricity but on darker days (when it rains) lights can be provided with a starter. Both Mr. Nkomo's and Mr. Mogale's learners are issued a textbook, although they work from worksheets most of the time. Mr. Peters' classroom has an "old style" desk with adjoined chair for each learner, which means that they cannot easily be arranged in groups of more than two. There is a teacher's desk and cupboard, a chalkboard and chalk, no overhead projector and screen and no electricity, so on rainy days the classroom is dark. Some windows are broken. The school has some textbooks but there are not enough for each learner and so they are not issued and Mr. Peters works from worksheets that he develops.

I have spent some time describing the schools and the classrooms. This serves three purposes. First, it gives the readers a picture of the contexts so that they might better understand our later analyses of teaching and learning in these classrooms. Second, it shows that the five schools in which we worked are generally better resourced than most South African schools, although not as well resourced as many classrooms in "developed" countries. Third, it shows the differences in resources and socio-economic profiles across the five schools, which will allow me to make some claims in relation to equity and teaching and learning mathematical reasoning. The five classrooms create a matrix design across grades and socio-economic status as described in Table 2.4. I will show later in the chapter how this design is both reinforced and complicated by learners' knowledge in relation to their socio-economic status and by the tasks chosen by the teachers.

Table 2.4 Variation across schools

Grade	Socio-economic status	
	High	Low
Grade 11	Mr. Daniels	Mr. Nkomo
		Mr. Mogale
Grade 10	Ms. King	Mr. Peters

Learner Knowledge

Learner knowledge was ascertained across the five classes through classroom observations and task-based interviews with learners. It was clear that Mr. Peters' learners had extremely weak knowledge, probably a few grade levels below Grade 10. Mr. Nkomo's learners were closer to grade level but showed some weaknesses, particularly in relation to Mr. Daniels' learners, who were around grade level. This is somewhat surprising, given that Mr. Nkomo's class was untracked and consisted of both higher and standard grade learners, while Mr. Daniels' class was a standard grade class only (see Table 2.2). The learners' knowledge in these cases reflects the socio-economic status of the schools. Further reflecting socio-economic status, Ms. Kings' Grade 10 learners were extremely strong and were the second highest class in a strongly tracked grade. In fact, their knowledge was stronger than both Mr. Nkomo's and Mr. Daniels' Grade 11 learners. Mr. Mogale's learners provide an interesting counterpoint to the SES/learner knowledge link. Although his is a low SES school, this class had been chosen in Grade 9 as the strongest learners in the grade, had been kept together as a class and had Mr. Mogale as their mathematics teacher since then. He had worked to build their mathematics knowledge and confidence over 3 years, informed by the principles of the new curriculum, which he had learned on in-service workshops.

In the task-based interviews, two or three learners were interviewed together and encouraged to help each other. High-achieving learners were chosen from Mr. Nkomo, Mr. Peters, and Mr. Daniels' classes while the learners from Mr. Mogale's and Ms. King's classes were close to average achievement.

Mr. Peters' learners showed very little facility in solving mathematics problems. Their solutions showed procedural errors in almost every step and suggested that they were looking for rules, rather than thinking about the meaning of what they were doing. Even the rules that they did remember, for example "what you do to the one side, you do to the other" were almost always applied incorrectly. Occasionally, with the easier problems, one learner used trial and error methods and obtained correct solutions but then did not know what to do with these, nor how to relate them to the mistaken rule-based calculations. Occasionally through prompting by the interviewer, this same learner was able to make some conceptual connections. The two learners did not manage to communicate with each other in ways that helped their problem solving; rather their conversations seemed to encourage even more mistakes and misunderstandings.

Mr. Nkomo's learners showed some facility with mathematical procedures and calculations and working algebraically without mistakes. However, they did not relate their calculations to the underlying mathematical meaning and when they were confronted with something slightly out of the ordinary, could not make sense of it. They did not use trial and error methods and were heavily dependent on the interviewer's help to solve most of the problems. They were able to work together and sometimes help each other with procedural issues.

The learners in Mr. Daniels' class showed procedural fluency and were able to talk conceptually about the mathematical solutions they were developing. They were able to reason with mathematical objects, although some of their reasoning was flawed and somewhat problematic from a mathematical standpoint. They spent useful time talking and explaining ideas to each other, correcting mistakes and resolving conflicting ideas, while checking that their procedures were correct and eventually coming to consensus on most solution methods. They were able to work with the interviewer's prompts and incorporate them into their own problem solving activities.

The learners in Ms. King's class were much more procedurally fluent with equations than both Mr. Nkomo's and Mr. Peters' learners, and even slightly more fluent than Mr. Daniels' learners even though they were a grade lower. They had been taught some Grade 11 concepts and procedures as "extras" in Grade 10 and were able to work with these as well, with some assistance from the interviewer. Conceptually, one learner struggled with some of the same issues that Mr. Daniels' Grade 11 learners struggled with, while the other was able to reason mathematically in a particularly perceptive way. The two boys were able to work together and help each other.

The learners in Mr. Mogale's class were procedurally fluent. They made occasional mistakes, which they noticed themselves because they continuously checked their work, looking for mistakes. They also estimated answers as a check on their procedures. They understood the meanings of the mathematical objects they worked with and reasoned mathematically with them. They went further than the learners in the other classes, in that they noticed links with other areas of mathematics and posed interesting questions about their observations. So they extended their thinking, creating new conjectures about the relationships between mathematical ideas.

These differences in the learners' knowledge were evident in the classroom interactions as well. These differences cannot be read as a comment on the particular teachers in this study (except possibly in the case of Mr. Mogale who had taught these learners for 3 years). It is clear that both the strengths and the weaknesses in the learners' knowledge comes from prior years of schooling and is a function of far more than only particular teachers. In four of the five classrooms, learners' knowledge is strongly associated with the racial and socio-economic profiles of the schools. This makes sense because schools in poorer areas usually have fewer resources, larger classes, and generally, less knowledgeable teachers (Fleisch 2007). The strong association of learner knowledge with race and socio-economic status in this sample modifies the matrix design in Table 2.4 slightly (see Table 2.5 below). Given that Mr. Mogale's learners provide a strong exception to the rule, being of low socio-economic status but with strong learner knowledge, we will be able to make some arguments which de-link learner knowledge and socio-economic status in subsequent chapters.

Table 2.5 Variation across schools

Learner knowledge/SES Grade	Strong/high	Weak/low
Grade 11	Mr. Daniels	Mr. Nkomo
	Mr. Mogale (knowledge)	Mr. Mogale (SES)
Grade 10	Ms. King	Mr. Peters

The Tasks

As part of the research design, the teachers worked together to plan tasks, which would engage learners in mathematical reasoning. The two Grade 10 teachers worked together and the three Grade 11 teachers worked together. They worked with drafts of new South African textbooks which were being developed in relation to the new curriculum (Jaffer and Johnson 2004; Johnson et al. 2006), as well as some of their own resources, in order to choose, modify, and develop tasks that they thought would be useful to elicit mathematical reasoning and would also fit in with their curriculum. They spent two sessions of 2½h each, planning the tasks and how they might teach them.

The Grade 11 Tasks

The tasks developed by the Grade 11 teachers (see Appendix) aimed to get the learners to explore how horizontal and vertical shifts of a parabola on a Cartesian plane produce differences in the equations of the graphs. The task consisted of three activities. The first activity required the learners to trace a copy of the graph $y=x^2$ onto a transparency, shift the transparency three units to the right and four units to the left, and observe what happened to the values of corresponding points on the shifted graphs in relation to the original graph. The second activity showed the original and the two shifted graphs on the plane, with their equations: $y=(x-3)^2$ and $y=(x+4)^2$ and asked learners to compare and contrast the graphs, and then to focus on the more general question of how the value of p in $y=a(x-p)^2$ affects the graph. The third activity dealt with vertical shifts, with the graphs of $y=(x-3)^2$, $y=(x-3)^2+2$ and $y=(x-3)^2-3$. Again the learners were asked to compare and contrast the graph and then answer the more general questions of how the value of q in $y=a(x-p)^2+q$ affects the graph.

An analysis of the task using Stein et al.'s (2000) framework (see Chap. 3 for more detail) shows that that it demands higher level thinking from learners, predominantly at level three – "procedures with connections". According to Stein and her colleagues' criteria, the activities suggest pathways, to follow, that are closely connected to underlying conceptual ideas, are represented in multiple ways to help learners build connections and develop meaning, and require learners to engage with conceptual ideas in order to complete the task successfully.

Two additional aspects of the task are important for the subsequent analysis. First, the task is inductive, in that it asks learners to explore particular examples of shifting graphs and then make generalizations based on these examples. It does not ask for any form of deductive proof or justification. Exploratory questions such as "Discuss with a partner how these graphs differ and are the same" and "What do you observe?" are relatively open and unconstrained, except by the graphs, in what could count as an acceptable response. Learners could comment on only one observation or on as many as they could find. They could comment with or without explicit justification.

Second, the task contains a number of possibilities for misconceptions to arise and become visible. A key, counter-intuitive idea that is entailed in the task is that when the graph shifts in the positive direction, the equation has a negative sign in the brackets, i.e., $y=(x-3)^2$ is the equation of the graph that shifts three units to the right. Similarly, when the graph shifts in the negative direction, the sign in the brackets becomes positive. Many learners in all three Grade 11 classrooms demonstrated the misconception that the sign in the brackets should follow the direction of shift of the graph, making for some interesting discussions and exploration of the links between equations and graphs. Other conceptual issues that learners struggled with were: what does it mean if a graph extends infinitely along one axis; what counts as corresponding points; and how to read variables such as p and q in an equation. How these misconceptions influenced the teaching of mathematical reasoning in these classes is discussed in subsequent chapters.

The Grade 10 Tasks

Ms. King and Mr. Peters began their planning by looking for tasks that would enable learners to engage in all five strands of mathematical proficiency identified by Kilpatrick et al. (2001): conceptual understanding; procedural fluency; strategic competence; adaptive reasoning and productive disposition (see Chap. 6 for more detail). Ms. King wanted to focus on the integrated development of all five strands among her learners while Mr. Peters wanted to focus on the adaptive reasoning strand and develop his learners' ability to justify their thinking. Given their slightly different foci, and because of Mr. Peters' concerns that the tasks might be too challenging for his learners, Ms. King and Mr. Peters used the same first task, but different subsequent tasks. I will first discuss Ms. King's tasks and then Mr. Peters' tasks (see Appendix).

Tasks 1 and 5 on Ms. King's worksheet are primarily deductive tasks in that they require the learners to evaluate conjectures as true or false and then justify their decision. Task 1 asks whether x^2+1 can equal zero and what the smallest value for x^2+1 is if x is a real number. Task 5 asks whether n^2-n+11 is a prime number, if n is a natural number. Learners might test the conjectures using specific examples. However, they do not need to, they could work on a general justification from the beginning. In the case of task 5, if they do test examples, the first 10 natural numbers will give prime numbers but 11 will not and so makes the point that inductive testing is not good enough because there can be counter-examples. In this case, the general argument is that for n^2-n+p, $n=p$ gives p^2 which is not a prime number. Tasks 1 and 5 make demands on learners who are at level four (the highest level) of Stein et al.'s (2000) framework, which they call "doing mathematics". The tasks require nonalgorithmic thinking, they do not suggest specific solution approaches, they require learners to integrate existing knowledge to form understandings of new relationships, they require learners to examine task constraints, and they require some self-regulation and self-monitoring of the learners' thinking processes.

Tasks 2 and 3 require learners to work with the definition and meaning of a function, and with function notation. They make level three demands in Stein and her colleagues' hierarchy, suggesting solution methods that connect to underlying meanings and requiring multiple representations. Task 4 gives learners practice with function notation, which Ms. King thought was important in helping learners develop both conceptual understanding of and procedural fluency with functions. This task can be enacted at either level two or three of Stein's hierarchy, depending on how individual learners approach it. The task can be approached using the procedures of substitution and simplification of algebraic expressions without thinking much more deeply about the notion of a function. Alternatively, the task might support learners to make connections with what they have done before, and come to a more fluent and better understanding of functions. Because making connections is not explicitly asked for in the task, this task would be considered to be a level two task – "procedures without connections".

Mr. Peters worked with the same first task as Ms. King, although he excluded the second part: what is the smallest value for $x^2 + 1$. His first task read: Consider the following conjecture: "$x^2 + 1$ can never be zero". Prove whether this statement is true or false if $x \in R$. During the planning process, Mr. Peters expressed concern about how his learners would approach this task because of their weak knowledge. He worried about moving on with the same tasks as Ms. King, expecting that his learners would need more time working with the first task and that he would need to give them additional guidance. He developed a second task (Task 1B) where he scaffolded the learners' substituting into various single-term expressions and working with the sign of the expression. After two lessons where learners struggled with Task 1, he decided that this task (1B) would not help, as learners tended to focus on the sign rather than the value of the expression. So while teaching and monitoring learners' responses, he developed a second task that he hoped would address these difficulties, because the sign in front of each expression is not the sole determining factor of the sign of the expression (Task 2).

Both Tasks 1 and 2 are primarily deductive and can be approached by using a combination of inductive and deductive methods. Mr. Peters hoped that both tasks would encourage the learners to use a combined inductive–deductive approach, through substituting, testing, and justifying conjectures. In Task 2, he had the more specific goals of developing procedural fluency in substituting into the expressions, conceptual understanding that the expressions represent a range of values, strategic competence in that learners should not read off whether the expression was positive or negative from superficial aspects of the expressions and adaptive reasoning in that they justified their answers. Task 1 would count as "doing mathematics" in Stein and her colleagues' hierarchy, while Task 2, as Mr Peters intended it to be solved, would count as "procedures with connections". In Task 2 there is a specified solution method (not in the task as such but Mr. Peters made it clear in class), which is intended to help learners make connections with underlying meanings and concepts.

The above discussion of the tasks has shown differences in the ways they support learners to make connections between procedures and meanings; integrate the various strands of mathematical proficiency; and how they constrain what might count as an acceptable solution. In subsequent chapters, we will show how the tasks afforded and constrained the contributions that learners made in the classroom and how the teachers dealt with these. For the purposes of this chapter, the discussion of tasks fills out the matrix design of the study in Tables 2.4 and 2.5. Given that the tasks used in each grade were the same or similar, comparisons within and across grades are made easier. The matrix design in Table 2.6 enables some extrication of the variables of task, learner knowledge, and socio-economic status in relation to the possibilities for teaching mathematical reasoning in differently resourced classrooms.

All grade 11 teachers used the same tasks, which were inductive and which supported procedures with connections to meaning. Mr. Daniels' learners had strong mathematical knowledge and were of high SES while Mr. Nkomo's learners had weak mathematical knowledge and were of low SES. Mr. Mogale's learners provide a contrast to the others in that they were of low SES but had strong mathematical knowledge. The two grade 10 teachers used similar tasks, which were mainly deductive and which varied from "doing mathematics", through procedures with connections to procedures without connections. Ms. King's learners were of high SES and had very strong mathematical knowledge while Mr. Peters' learners were of low SES and had weak mathematical knowledge. These similarities and differences among the teachers enable comparisons in relation to tasks, school context, and learner knowledge, which we pursue in Part 3 of this book. In Part 2, we look at the individual case studies of each of the teachers, which provide in-depth descriptions of their teaching of mathematical reasoning.

Table 2.6 Variation across teachers in tasks, learner knowledge and SES

Learner knowledge/SES Tasks	Stronger/higher	Weaker/lower
Grade 11 Inductive Procedures with connections	Mr. Daniels Mr Mogale (knowledge)	Mr. Nkomo Mr Mogale (SES)
Grade 10 Deductive (with some inductive) Procedures with and without connections, doing mathematics	Ms. King	Mr. Peters

Introduction to Part 2

The next five chapters each present a case study of one of the teacher's classrooms. All of the teachers were inspired by the same general concern – how to develop the teaching of mathematical reasoning as encouraged by the new South African curriculum in their particular context. This general concern was translated into more specific research foci by each of the teachers, each of which illuminates a specific aspect of teaching mathematical reasoning and which, taken together deepen our understanding of what such teaching demands of teachers and learners. These chapters, individually and collectively help to address some of the challenges identified in Chap. 1:

- How do learners respond to tasks chosen to elicit their mathematical reasoning?
- How can teachers interact with learners around tasks to engage their mathematical reasoning?
- How can teachers teach to develop mathematical proficiency?
- How does collaborative conversation among learners and teacher promote mathematical reasoning?
- What kinds of teaching practices, questions and moves help to encourage and sustain learners' mathematical reasoning?

Chapters 3–5 focus on the Grade 11 teachers and learners. In Chap. 3, we look at learners' responses to tasks that require making connections between procedures and meanings. We show that the learners, with weak mathematical knowledge, struggled to respond to these tasks, which in turn challenged the teacher to respond in ways that would deepen learners' engagement with the tasks. The teacher provides an honest reflection of how he struggled to do this in the beginning, improving as the week continued. In Chap. 4 ,we look at the learners' engagement with the same tasks. In this classroom, learners had stronger mathematical knowledge, and were able to engage more deeply with the task demands. We show how a collaborative conversation among learners and the teacher supported the learning trajectory of one learner, as he or she deepened his or her engagement with the task and with mathematical reasoning. The learning trajectory provides a description of how mathematical reasoning can develop among learners and what learners can achieve. In Chap. 5, we focus more closely on the teacher's practices and moves and show

how these were internalized by the learners to support their mathematical reasoning. These learners, also with stronger mathematical knowledge, were able to use some of the teacher's moves to support both their own and each other's reasoning.

Chapters 6 and 7 focus on the Grade 10 teachers and learners. In Chap. 6 we look at the extent to which teaching planned to include all five strands of mathematical proficiency, managed to achieve these. In Chap. 7, we focus on the strand of adaptive reasoning and the mathematical practices of justification and explanation. Again, we see that learners with weaker knowledge struggled to engage with the tasks and we see the teacher's reflections on the challenges that he or she experienced in trying to engage the learners in conversation.

In the case studies, all of the teachers draw on Kilpatrick et al. (2001) notion of mathematical proficiency and Ball (2003) notion of mathematical practices. These are discussed in detail in Chap. 1 and in each of the teacher's chapters, and so I will not repeat them here. A second general concern is teacher–learner interaction and the notion of "facilitator" as described by the South African curriculum documents. The notion of facilitator is informed by constructivism, which asserts that knowledge is not received by learners in the same form as transmitted by teachers, rather it is interpreted and restructured in relation to the learner's current knowledge. This introduces uncertainty into the teaching–learning process because teachers cannot know what learners have learned, although they can infer this from learners' contributions. This argument is sometimes taken to mean that the teacher should reduce involvement in teaching and should not "tell" (Chazan and Ball 1999) or give learners mathematical information, in case they interfere with learners' ability to construct the knowledge for themselves.

This last point is one that has been taken up by many South African teachers in ways that may be counterproductive for teaching and learning, as they give little or no direct mathematical input, particularly when learners work in groups. Such non-intervention was not an option for the teachers in this study; they were more interested in finding ways to intervene more appropriately. They drew on a Vygotskian perspective, which asserts that the teacher's input has a marked effect on learners' constructions in both positive and negative ways; teachers' responses can both support and inhibit learner constructions. We show how the five teachers influence learning in far more complex ways than the easy aphorism that too much telling inhibits learning. The teachers in this book walk a fine line between telling and not telling; questioning and not questioning; being silent and not being silent. Each of these can be appropriate at particular moments in the teaching–learning process, and moreover, it is the quality of the response that counts, not necessarily the kind of response. I have argued elsewhere (Brodie 2007a) that the dominant mode of question-and-answer methods, preferred by many South African teachers, does not necessarily make for more and better participation among learners. The chapters that follow provide alternatives to the question-and-answer method, which are both less and more directive, and we believe, more appropriate.

One teaching practice that is often ignored is that of listening (Davis 1997). The reason why many question and answer exchanges do not support mathematical reasoning is that teachers do not listen carefully enough to learners' contributions in order to formulate their next questions. The questions are driven by the teacher's pre–determined agenda, rather than by the need to respond to and develop learners' understandings (Davis 1997; Nystrand et al. 1997). All of the case studies that follow show teachers listening more carefully to learners' reasoning before making decisions on how to follow up on learner contributions.

Finally, all of the teachers used tasks of higher cognitive levels in order to elicit and engage with learners' mathematical reasoning. I argued in Chap. 1 that while tasks are necessary to develop mathematical reasoning, they are not sufficient; teacher–learner interactions of particular kinds and qualities, as discussed above, are equally necessary. What we add to the picture in these chapters is that tasks and interactions are experienced differently by different learners, and hence their teachers, particularly if their teachers are responsive. We show that in classes where learners have weaker mathematical knowledge, teaching mathematical reasoning is harder to achieve. This is of serious concern to those of us concerned with equity because if learners who are already advantaged become more advantaged through teaching methods, which they are better positioned to respond to, then the gap will widen rather than narrow. These case studies, particularly Chaps. 3 and 7 do suggest ways in which teachers of weaker learners can make progress, and also that they need support to do so.

We note here that this work took place prior to the introduction of the new curriculum in South Africa in Grades 10 and 11, which are the grades we worked in1[1]. However, the draft of the new curriculum was available, which is the document that the teachers worked with. At the same time, all of the teachers were working with the new curriculum in Grades 8 and 9, which has similar principles of engaging learners in mathematical tasks and conversations. The teachers had different experiences with new curriculum ideas, for example Mr. Mogale (Chap. 5) and Mr. Daniels (Chap. 4), who had been on lengthy in-service programmes, had been trying out new ideas for some time, even before the introduction of the new curriculum in Grades 8 and 9. Ms. King (Chap. 6) had been involved in a local project in her school, which worked with similar principles. Mr. Nkomo (Chap. 3) and Mr. Peters (Chap. 7) had much less experience than the other three teachers and were trying out some ideas for the first time. None of the teachers viewed themselves as accomplished "reform" teachers, and all saw this as an opportunity to confront old and new challenges and work with them. None of the learners had experienced this kind of teaching before, except in these teachers' classes, and so were inexperienced "reform learners", who still had to learn the norms of such teaching (McClain and Cobb 2001) as the teachers point out in the case studies.

As noted in the introduction, all of the five teachers were enrolled on an Honours Programme[2], which had addressed a range of issues, including broadened conceptions of mathematics and mathematics learning, mathematical reasoning, language

---

[1] The new curriculum has only recently been implemented in Grade 10 (2006) and Grade 11 (2007).
[2] In the South African Higher Education system an Honours degree follows a 3-year undergraduate degree and a professional teaching qualification and is necessary for entry into Masters.

and mathematics learning, and formative assessment. They were therefore better informed about the mathematics-specific issues in the new curriculum than many South African teachers, who have experienced generic, rather than subject-related training, and whose understandings of the new ideas are relatively superficial (Chisholm et al. 2000). Each teacher conducted her/his study as an Honours research project, and therefore these chapters can also provide ideas and guidance for teacher research projects conducted at this level.

Chapter 3
Mathematical Reasoning Through Tasks: Learners' Responses

In this chapter, I draw on the notion of mathematical reasoning discussed in Chap. 1 as well as the approach to mathematics in the new curriculum in South Africa. The new curriculum takes the view that mathematics should make sense to all learners and that learners should be given opportunities to solve problems, look for patterns, make conjectures, make inferences from data, explain and justify their ideas and challenge others' ideas. The National Curriculum Statement Grades 10–12 (Department of Education 2003, pp. 9–10) describes the purpose of mathematics in the new curriculum as follows:
Mathematics will enable learners to:

- Communicate appropriately by using descriptions in words, graphs, symbols, tables, and diagrams
- Use mathematical process skills to identify, pose, and solve problems creatively and critically
- Organize, interpret, and manage authentic activities in substantial mathematical ways that demonstrate responsibility and sensitivity to personal and broader societal concerns
- Work collaboratively in teams and groups to enhance mathematical understanding
- Collect, analyse, and organize quantitative data to evaluate and critique conclusions arrived at
- Engage responsibly in quantitative arguments relating to local, national, and global issues

These broad curriculum outcomes resonate with the description of mathematical reasoning given in Chap. 1 and show the intention of the new curriculum to produce learners who can work with mathematics in a variety of ways. The intention has moved away from learners who use calculations and formulas as the only ways of solving mathematical problems, and producing only accepted correct solutions to problems. When our current learners leave school they will be facing a world different from that of their parents and teachers. They will need the mathematical skills of reasoning and justification to respond to a range of challenges. We do not help our learners to rise to these challenges by teaching them that all problems can be solved merely through the application of certain procedures. The National Curriculum

K. Brodie, *Teaching Mathematical Reasoning in Secondary School Classrooms*,
DOI 10.1007/978-0-387-09742-8_3, © Springer Science+Business Media, LLC 2010

Statement argues that there are many ways that learners can learn mathematics successfully. Learners should be given opportunities to communicate their ideas critically and creatively as they work on mathematics tasks.

One of the key influences in how learners learn to reason mathematically is the nature of the tasks that they work with in class (Stein et al. 1996, p. 72, 2000). In this chapter, I analyse Grade 11 learners' work on tasks that involve mathematical reasoning. At the time of the study, the Grade 11 syllabus did not require learners to solve non-routine problems that they had not seen before and did not include tasks that required learners to display their reasoning, or to formulate, test, and justify conjectures. The tasks mostly required learners to carry out calculations or procedures that they had been taught. In my experience during the past 10 years of marking Grade 12 examination papers, I had come to realize that learners perform poorly on tasks that involve higher order mathematical reasoning. This chapter aims to explore the challenges that learners may encounter with tasks that involve mathematical reasoning and the ways in which a teacher can help learners to improve their mathematical reasoning.

Tasks that Support Mathematical Reasoning

Mathematical tasks are given to learners by the teacher to engage them in mathematical activity in order to develop certain mathematical concepts or practices. Stein et al. (1996) defined a mathematical task as a classroom activity, which is intended to focus learners' attention on a particular mathematical idea. Once the teacher has set up learning goals, s/he can give tasks that match with her/his goals for the kinds of thinking s/he would like the learners to engage in. If the teacher wants learners to memorize mathematical facts and procedures she/he will give tasks that require memorizing. The old curriculum tended to prioritize memorizing over other forms of mathematical activity and most textbook and examination tasks required learners to memorize and recall facts and procedures. The new curriculum requires a broader range of mathematical practices and if teachers are to help learners develop these, we will need to broaden the range of tasks that we ask learners to engage in.

Stein et al. (2000) distinguished between two levels of cognitive demand of mathematical tasks, and within each of these, two kinds of tasks. Lower level tasks are memorisation tasks and tasks that require procedures without connections to meaning or concepts. Higher-level tasks are those that require procedures with connections to meaning or concepts and "doing mathematics" tasks, which require a high level of exploration from learners. These are described below.

Memorization tasks involve reproducing previously learned facts, rules, formulae, or definitions. Memorization tasks do not require any explanation from learners; they are straightforward and learners use well-known facts to solve them. *Procedures without connection tasks* require reproduction of procedures but without connection to underlying concepts. Such tasks are focused on producing correct solutions rather than developing mathematical understanding. *Procedures with connection tasks* focus learners' attention on the use of procedures for the purpose of developing deeper levels of understanding of mathematical concepts and ideas. Such tasks focus learners on the procedure of solving mathematics problems in a meaningful way. *Doing mathematics tasks* do not require any procedure to be followed. There is no predictable way of solving these problems. Learners working with such tasks need to analyse task constraints and creatively find their own solutions.

Since learners interact with tasks, the cognitive demands of the tasks depend not only on the task, but also on the learner. For example, a task that is at a high level for a Grade 8 learner might be at a lower level for a Grade 11 learner; or two learners in the same grade might solve the same task in different ways. One might use procedures and think about the connections of the procedures to the underlying mathematical concepts, while another might use the same procedures without making any connections to meaning or concepts.

The cognitive demand of the tasks can be recognized in the task features, which include the "number of solution strategies, number and kind of representations and communication requirements" (Stein et al. 1996, p. 455). Tasks that can be solved by using different approaches, and those that require learners to bring together different representations and to explain, justify, and communicate their ideas are likely to be of higher cognitive demand than those that have only one method of solution and do not require additional effort in working with representations and explaining ideas. In order to encourage mathematical reasoning as suggested by the new curriculum, teachers should give learners tasks that allow for different levels of engagement, including the higher levels. If learners are given only tasks that are of a lower level, they will find it difficult to tackle higher-level tasks.

Examples of lower-level tasks are found in many textbooks, for example "Sketch the graph of $y=(x-3)^2+2$". In response to such tasks most learners will use the usual procedure of finding the x and y intercepts, the turning point and draw the graph. Most learners use these procedures without understanding the relationship between the equation and the graph, however as mentioned above, some (very few) learners do make connections with the meaning of the graph. An example of a higher-level task from a new curriculum textbook (Bennie 2006), explicitly asks learners to make connections between procedures and meanings and to justify conjectures:

(a) Draw a sketch of $y=x^2$. Use a table if necessary.
(b) Consider the graphs with equations $y=ax^2$. Make a conjecture about the effect on the graph when you change the value of a. Test your conjecture by drawing the graphs of $y=2x^2$, $y=3x^2$, and $y=1/2x^2$ on the same system of axes you used in (a).
(c) What will happen if the value of a in $y=ax^2$ is negative? Choose suitable values for a and test your conjecture.
(d) Summarize your observations in questions a and b above, etc.

This task requires learners to go beyond drawing the graph and to make conjectures from the graph. It also requires them to summarize their observations, which is something that was not visible in the old curriculum.

Stein et al. (1996, 2000) distinguish between two phases of task-use in the classroom: task set-up and task implementation. They note that the task demands can shift between the set-up and implementation phases, depending on how learners engage with the task and how teachers interact with learners. Often tasks that are set up at a higher level, decrease in level as they are implemented (Modau and Brodie 2008; Stein et al. 1996, 2000). This can happen for a number of reasons: learners might choose to work at a lower level than the task requires, ignoring the higher level task demands; and teachers might give learners too much help, which reduces the level, for example "funnelling" the task (Bauersfeld 1988). Stein et al. (1996) argue that "classroom norms, task conditions and teachers' and students habits and dispositions" (p. 461) can all influence how tasks change at implementation. Their research shows that tasks are usually maintained at the same level or decline in level. The higher-level tasks in particular tended to decline.

Teaching for Mathematical Reasoning

Stein et al.'s task framework is consistent with both constructivist and socio-cultural perspectives on learning, which were discussed in Chap. 1. The notion of cognitive demand suggests individual or group engagement with tasks that promote learning, thinking, and reasoning at different levels in the individual. The idea that the level of the task changes in interaction between teacher and learner, or among learners suggests a socio-cultural perspective, because social aspects of the classroom determine the level of individual cognition.

These theoretical frameworks suggest important implications for teachers. Constructivism shows how errors and misconceptions can be deep-seated in learners' conceptual structures and learners have to do the difficult work of transforming their own thinking, with their teacher's help. If learners' responses are incorrect, it does not necessarily help to tell them so. As Heaton (2000) remarks: "telling him he was wrong would not necessarily change how he thought". Heaton also goes on to say that if learning and teaching are about understanding why answers are right or wrong, then it is important to explore learners' errors with them. The teacher can ask learners to explain their answers by posing questions like: "can you explain how you got the answer" or "convince me that your answer is correct". Moreover, asking learners to explain is just as important when their answers are correct as when they are incorrect because explaining can deepen their own thinking, and help others in the class. It also encourages the "social norm" (Yackel and Cobb 1996) that all answers should be justified in mathematics.

A socio-cultural perspective suggests that by asking for explanation and justification, the teacher can make learners aware of their reasoning and support them to construct appropriate mathematical ideas. The teacher can, through questions and prompts, try to provoke learners into thinking in particular ways and support them to compare, verify, explain, and justify their conjectures. It is not easy for the teacher to ask questions that will make learners aware of their reasoning, since learners might not respond as the teacher expects. Ball (2003) argues that when a teacher whose students have never been asked to explain their thinking asks them to justify their solution, s/he is likely to be greeted with silence. When s/he asks a learner to explain her/his method, the learner will probably think that she/he made an error. It is therefore important, but demanding, for teachers to develop norms of interaction in their classrooms (Yackel and Cobb 1996, see Chap. 1 for more discussion).

So teaching for mathematical reasoning involves teachers being aware of the type of tasks that we give to learners, choosing tasks that enable learners to make sense of mathematics and that give them opportunities to investigate, analyse, explain, conjecture and justify their thinking, and interacting with learners around the tasks to maintain, or even raise, the level of the task.

The Classroom and the Tasks

For the purposes of linking this chapter with chapter two, it is important to note that my pseudonym in the study is Mr Nkomo. This study was conducted in one of my Grade 11 classes, in a functional township school west of Johannesburg, with very basic facilities (see Chap. 2 for more detail). All the teachers and learners in the school were "black African" South Africans (see Chap. 2). English is not the main language of any of us, but all teaching and learning of mathematics occurs in English. There were 1,650 learners in the school at the time of the study and 42 teachers, giving a teacher–learner ratio of 1:39. There were 28 learners in my Grade 11 class. Since fewer learners take mathematics in the higher grades, these classes tend to be smaller. The learners worked in ten groups of between 2 and 4 learners in a group. These groups were established prior to the study and learners were used to working with each other. The class was of mixed ability in mathematics and learners in the class were taking mathematics on both higher grade/standard grade.[1]

I planned a series of tasks on functions with the two other Grade 11 teachers as part of this project (see Chap. 2 and Appendix). The tasks were planned to take 1 week of class time and were intended to engage learners' mathematical reasoning as

---

[1] As explained in Chap. 2, at the time of the study, the Grade 12 mathematics examination could be taken at two levels, standard and higher grade. Success on the higher grade (or exceptionally good marks on the standard grade) granted access to scientific fields at university.

discussed below. As learners worked in their groups, I went around asking questions where necessary. I also conducted whole class discussions after each task. At the end of each lesson, I took in each group's work and read it carefully, preparing how to conduct a whole class discussion on different groups' work the next day.

In the next section, I will analyse the first task and so I describe it here (see Appendix for actual task). Learners were presented with the graph of $y=x^2$, asked to move it first 3 units to the right and then 4 units to the left. In each case, they were asked to compare the turning points of the two graphs ($y=x^2$ and each shifted graph), and after that to compare corresponding points on the two graphs. The task encouraged learners to experiment with shifting graphs, and to focus on particular points in order to begin to notice relationships between them. Since they were explicitly told how to do this, i.e. told first to look at the turning point and then at other points, and since they were given a table with values for $y=x^2$ filled in, which they had to complete, this is a "procedures with connections" task in Stein et al.'s (2000) framework. Learners were asked to write down their observations about how the points shifted as the graphs shifted and in doing so to make connections between the shifting graphs and shifting values of co-ordinates. Learners had to examine parameters, experiment with shifting graphs, make conjectures and write down observations.

Learners' Responses: An Overview

My first step was to analyse the learners' written responses to the tasks in three categories:

1. Comparing the turning point of $y=x^2$ to the turning point of the new graphs
2. Choosing and comparing other corresponding points on the graph
3. Observations about the corresponding points of the two graphs

Table 3.1 shows the total number of groups who gave correct and incorrect responses for each of the above (there were ten groups).

Table 3.1 shows that learners were able to identify and compare the new turning points of the graphs, but struggled to identify, compare and make observations about other corresponding points on the graphs. However, as discussed above, correct and incorrect responses are not sufficient to illuminate or develop learners' mathematical reasoning. So I continued to analyse learners' reasoning, both from the written tasks that they handed in, and from videotapes of whole class discussions.

Table 3.1 Correct and incorrect responses

	Category 1	Category 2	Category 3
Correct	10	1	0
Incorrect	0	9	10

Learners' Responses: Detailed Analysis

In the first part of the task (Category 1), learners were asked to compare the turning points of the shifted graphs to $y=x^2$ (the graphs were shifted 3 units to the right and 4 units to the left). The aim of this question was to help learners to connect the turning point to the position of the graph and to notice how the turning point changes as the graph shifts. As shown above, all the groups correctly identified the turning points of the shifted graphs. The only difference in the answers of the groups was that while most groups acknowledged that the turning point has two co-ordinates, and wrote the turning points as (0,0), (3,0), and (−4,0), two groups wrote only about the x-coordinate of the turning point, for example: "For the first graph the turning point was 0. After moving it 3 units to the right we observed the new turning point, which is 3". Although in some classrooms this might be considered to be "incorrect," because the y-coordinate of the turning point has been left out, in fact in this context, the learners had commented on the parameter that had changed because of the shifting graph, the x-coordinate. So these learners have reasoned appropriately for this task.

From the above, it is clear that this part of the task does not demand high levels of reasoning from learners because it merely requires them to identify a point. Although it might support some learners to connect the shifting point to the shifting graph, this part of the task does not explicitly require these connections and many learners may not have made them.

For the next part of the task (Category 2), choosing and comparing other corresponding points, only one group (Group 9) correctly completed the task. This group correctly recorded new values for x on each graph for the given values of y on the table. Their observation was that the x values increase for the first shift (3 units to the right) and the x values decrease for the second shift (4 units to the left). However, they did not give the exact value of the shift in either case. Another group (Group 10) worked from the x-values and managed to record correct values for positive x values for the shift to the right. This is what they wrote:

x Values	−3	−2	−1	0	1	2	3
y Values	Are not to be found			9	4	1	0

They did not give their reasons for saying they could not find the y-values for negative x-values. This response seems strange to me because they were physically shifting the graph using a transparency, and therefore I am not sure how they could not see the left hand side of the graph. It might be that this group did not understand the meaning of "corresponding points" because, even though they drew the above table, in the rest of the answer they only spoke about differences in x- and y-intercepts, rather than for a range of corresponding points. Only one other group (Group 6) tried to record values in a table and they did it incorrectly, and were unable to make a reasonable conclusion about the shifts.

The responses to this question suggest that learners struggled to make use of tables to understand shifting points on the graphs. It is very rare in the current Grade 11 textbooks to find tasks where learners are asked to record their findings.

It is also the case that they only use tables when initially introduced to graphs, so that they may not see them as a useful vehicle to think about graphs and solve problems with them. This task attempted to engage learners in making connections between the graphs and the points on them, and to move between different representations, definitely a "procedures with connections" task. The fact that so few learners were able to manage it suggests that higher-level tasks and working with different representations are difficult for them.

The learners' observations about the shifts in the points (Category 3) show some interesting patterns of reasoning, even among groups that did not record values in the tables. The different responses that I identified are:

(A) The x- and/or y-values change
(B) The x values increase/decrease
(C) The x-values are all positive/negative

Table 3.2 shows all the groups responses to this question, with the response labelled A, B, or C according to the above

Table 3.2 Groups responses to question 3

	Shifting 3 units right	Shifting 4 units left
Group 1	It has positive values. All values of x and y have changed (A and C)	All values of x are negative and all values of y are positive (C). In the middle and the left graph we find negative and positive values and the right graph has only positive values (C)
Group 2	In the new graph we have only positive x and y values (C)	The numbers which were negative become positive, when we move the graph to the left 4 units (C)
Group 3	On the new graph the x-values increase (B)	The x values decrease on the left graph because it includes the negative numbers (B and C)
Group 4	The x-values of the old graph has negative units and the new graph does not have negative units (C)	y-Values and x-values change again (A)
Group 5	All the points changed. The negative points became positive points (C)	The y-values and x-values changed (A)
Group 6	We did not have negative values in the new graph. When the turning point changes, the other points change (C)	
Group 7	The y-values are still the same but the x-values have changed from the first graph (A)	The graph is not in touch with the y-axis
Group 8	The table points are totally different and the points of the second graph are all positive (C)	The x-values have changed to be negative as the graph was moved to the left (C)
Group 9	x-Value increases, y-values do not change (B)	x-Value decreases, y-values do not change (B)
Group 10	Wrote about intersection points of two parabolas rather than comparing them	

Only two responses were not classifiable according to A, B, and C. Group 10's response, which was actually not a response to this task and so will not be discussed here, and Group 7's response for the second shift, which is interesting and will be discussed. Of the groups that made response type A, all except one claimed that both x- and y-values changed. This suggests that they were not looking at corresponding points, but at the graphs more generally. Group 7 made the correct claim that for the same y-values, the x-values change. However, they did not specify the direction or the nature of the change. So while this answer is correct, it suggests that the learners were not looking closely at the graphs to make the connections that the tasks expected. Groups 3 and 9 made response type B. Group 9 argued that the x-values would increase or decrease (depending on the graph), when the y-values remained constant. Group 3 did not mention the y-values but we can infer that they were looking at corresponding points on the graphs, which means that the y-values stayed the same. Neither of these groups indicated the magnitude of the changes (3 in one direction, 4 in the other).

Response type C, made by most of the groups, indicates an interesting misconception. As the learners' shifted the graph, it seemed that all the x-values moved into the first quadrant (for the first shift) and into the second quadrant (for the second shift). This would in fact have happened with the traced graph in the second, but not the first shift, because the graph does not show coordinates beyond $x=3.5$. However, the learners' responses in both cases suggest that they did not understand that these graphs extend to cover all x-values; they thought that the graph ends where the picture ends. This misconception was found in both the other Grade 11 classes (see Chap. 8). Group 7's response is also a version of this response in that they looked at the actual drawn graph and saw that the drawing did not cut the y-axis. They too did not realize that the graph must extend to cover the domain of all x-values.

These three response-types show up interesting issues in the learners' reasoning and in the task demands. The "most correct" of the three sets of responses is type B because there are no errors. However, these responses are not specific about the magnitude of the change. Since the task was quite open, and asked learners "what do you observe", this is a perfectly acceptable response. However, it suggests that learners did not ask themselves whether they could go further to observe anything more. In this case, I argue, the level of the task declined from procedures with connections at set-up to procedures without connections in the learners' responses. Response type A can also be correct, when it focuses on the corresponding points. In these cases, it is even less specific than response type B, because it does not note the direction of change. Here too, the learners did not push themselves further, but did write down what they observed, also signalling a task decline. Response type C is the most specific; it attempts to get to grips with the picture as well as the values and to go deeper into what is happening. In this case, the learners preserved the level of the task; they were looking for connections and trying to make a generalization from the picture, and in doing this, they revealed an interesting misconception in their thinking, i.e. they did not think that a parabola extends infinitely along the *x*-axis in both directions.

This task analysis has suggested that when the task was of a lower level, learners were able to get it correct. When the levels increased, learners struggled both to get the tasks correct and to maintain the higher levels of the task. Ironically, the one time when they did maintain a higher level, a misconception arose. In the next section, I look at what happened when I discussed the responses in a whole-class discussion with the learners.

Teacher–Learner Interactions

As both teacher and researcher, I needed to understand learners' reasoning within and beyond their written responses. Were they saying more than what they wrote? Did they have more to offer than their responses on the page? Did I interpret their responses correctly? Did the learners understand their own thinking? Finally I wanted the whole class to understand what other learners were thinking and reasoning. In this section, I will discuss four kinds of teaching interventions that I used to address the above questions. These interventions were drawn from the literature (Chazan and Ball 1999; Heaton 2000; Lampert 2001) and seemed to make sense for my classroom. They are: encouraging learners to participate and listen; using learners' contributions to move forward; and pushing for more explanation from learners. I show how these interventions became a learning process for me, at first I was not very successful in generating discussion and thinking around learners' responses but as I persevered, I got better at it.

Encouraging Participation

At the beginning of the next lesson I told the class that I had read their responses the previous night and I found them all interesting. I then told them that I would ask them questions based on what I did not understand in their responses. This was a new norm for the class because they usually assumed that I did understand and asked questions to hear their answers, a "testing" function of questions (Nystrand and Gamoran 1991). This was the first interaction of the lesson after the groupwork, and involved Themba from Group 3. I chose to start with this one because, similarly to the other groups, Group 3 had mentioned that the *x*-values increased but had not been specific about the size of the increase.

Mr Nkomo:	So, Themba, what, what did you mean when you say, the, the values of, the values of, the x values increases?
Themba:	You move the graph three units to the right. You'll find that three is the turning point, and then one, eh nought, one, two, they are greater than minus one, minus two, minus three because in the first graph, minus one, minus two, minus three were included.
Mr Nkomo:	Fine. Thank you Themba, Did you hear what Themba said?
Learners:	Yes.
Mr Nkomo:	Yes
Learners:	Yes.
Mr Nkomo:	Can someone say that again because you have just said yes.
Learners:	(*Silence*)
Mr Nkomo:	Hmm. You have said yes, Themba, would you say that again? They have said yes, but they don't want to say that, okay? Themba please say that, for the last time?
Themba:	First you move three units to the right. o finde [you find], the turning point is three, and the numbers including is nought, one, two, three, four, five, six. And then mo [in the] first graph, the numbers that were included were minus three, minus two, minus one.

The other groups did not respond to Themba's explanation and I was forced to ask him to explain again. So although I had hoped to generate discussion through Themba's response, I was not successful even in getting learners to repeat his idea. This was both similar and different to Heaton's (2000) experiences in trying out this kind of teaching for the first time. It is similar in that it is not very successful. It is different in that Heaton got responses from the learners but did not know what to do with them. In this case, I could not even get responses from the learners. It might be that because I was asking for an explanation, and they were not used to listening to each other's explanations, they felt uncomfortable about repeating it to the class. This suggests that it is important to build learners' listening skills, as well as their abilities to explain each other's ideas (Lampert 2001). In the case of this interaction, the level of the task declined – from one where procedures with connections were expected to one where connections were not made.

Using the Contribution to Move Forward

The next extract comes soon after the previous one. Here I focus on the responses that indicated that y-values stayed the same. I was trying to focus the class onto the notion of corresponding points.

Mr Nkomo:	Jacob's group said that the x-values change but the y-values don't change, they stay the same, do you agree with them? Jabu, can you say more about it. Can you come and maybe explain why they are still the same?
Jabu:	(*Comes up to the board*) Eh! On the new graph, there is still one and on the old graph, there is one. Four and four, nine and nine (*he points on the graph*).
Mr Nkomo:	Nine and nine?

Jabu:	Yes.
Mr Nkomo:	So, in other words, eh, you are saying, there is still one and one?
Jabu:	Yes.
Mr Nkomo:	There is still, once again?
Jabu:	Four and four, and nine and nine.
Mr Nkomo:	Four and four, and nine and nine.

In this case, I asked a particular learner to come up and explain another learners' contribution, rather than leave it up to the whole class as previously. Jabu was not in Jacob's group and his group's response had been different, so I asked him to come and explain Jacob's point. Here I was enabling broader participation and also teaching learners that they could respond to each other's ideas and explain them. This is what had gone wrong in the first interaction. Chazan and Ball (1999), in discussing the teacher's role in discussion-intensive teaching, suggest that while "telling" is not often a good idea for the teacher, what is important is to decide what to do instead of telling. In this case, I asked one learner to explain another learner's idea and that helped the lesson to move forward (Heaton 2000). However, in this case, the interaction still remained at a lower level, as I did not push the learners to really explain and justify their thinking.

Pushing for Explanation of Particular Ideas

In the following extract, I deal with the response type C identified above through asking Group 1 to explain their answer. I first repeated it for the class.

Mr Nkomo:	Kefilwe's group said that it has positive values, all y and x has changed. In the middle and left graph we find negative and positive values and the right graph has only positive values.
Kefilwe:	You see, pointing at the graph that was moved to the right, it has positive values and the other in the middle has positive and negative while the left has positive and negative.
Mr Nkomo:	(*Kept quiet, looked at the whole class and saw Meshack shaking his head.*) Meshack, it seem you disagree with what Kefilwe said, can you tell us why?
Meshack:	Sir, it doesn't mean that always when we move a graph to the right it will have positive values only and to the left negative values only. Sir, for example if we move the graph that was moved four units to the left, two units to the right it will still have negative x-values.
Mr Nkomo:	What about the one moved to the right?
Neo:	If we move it one unit to the left it will still have positive values.
Mr Nkomo:	What can you conclude from what Meshack and Neo said?
Lydia:	I think it depends on how far you move the graph, so it doesn't mean to the left negative values and to the right positive values.

I realized that Kefilwe's group had a misconception about the x-values of the shifted graph. What I was doing here was "asking a question for the purpose of helping her see something new, not to merely share her ideas" (Heaton 2000). Calling on Meshack was a deliberate strategy to help Kefilwe's group see what they needed. In this extract, I was also more successful in getting more learners involved. Once Meshack had made his contribution, Neo echoed it in relation to the other graph and Lydia was able to draw the two contributions together and summarize them. It should be obvious to the reader that I was far less present in this extract than in the previous ones; however, since my questions were clear and focussed (Watson and Mason 1998), it was a more successful intervention. I was able to get learners to explain their own ideas and listen and build on the ideas of others (Heaton 2000). Here we managed to maintain the task demands at a higher level, procedures with connections to meaning.

Unfortunately, the above interchange did not quite solve the learners' problem. While they seemed to agree that shifting to the left and right will not always create positive and negative x-values, they did not quite agree that it did not in the two cases under discussion. In fact, they may have only agreed for the cases that were moved fewer than 3 units to the right and 4 units to the left. So, there was still work to be done on this misconception.

Conclusions and Implications

There was a great deal of evidence in my data to suggest that learners had difficulties in responding to tasks that involve mathematical reasoning. Learners had several difficulties in responding to the more demanding parts of the task. These included not understanding what corresponding points were and why they should be compared, making limited observations about the corresponding points of the graphs, and not recording their observations in the table provided. It was evident that the learners were not used to such questions and did not know how to approach them. The learners were not used to "exploring" as much as they can in tasks. However, there was also evidence to suggest that in some cases, some learners did go deeper, and did maintain the connections to meaning, although they did this incorrectly.

My interactions with learners did not manage entirely to shift the levels of the task. I was not able to help the learners to bring the level of the task back to that of procedures with connections, except in the last case, where they had been making the connections. However, my analysis of my teaching suggested three very different kinds of interaction, which improved over the course of the week and finally did put the learners in conversation with each other. My analysis confirms Heaton's (2000) argument as to how difficult the task of engaging learners in genuine discussion about reasoning really is.

In undertaking this project, I worked with tremendous support – with my four colleagues and with my supervisor. We had all read a number of books and articles about teaching, helped each other to prepare and supported each other in analysing our lessons. Given all this support, the fact that I still experienced difficulties suggests that the ideals of the new curriculum will not be easy to achieve. At the same time, however, in trying out these ideas, I have learned a tremendous amount, which will improve my teaching from now on. I have learned to:

- Be aware of different levels that are entailed in tasks and to give learners a wider variety of tasks
- Accept learners' responses, whether right or wrong
- Assist learners by asking focused questions to let them see something new rather than telling them the correct answer
- Listen to what learners are saying during discussion to have a clearer meaning of their understanding
- Follow up learners' written work with discussion of their meanings in class
- Work harder to support learners to justify their mathematical ideas

I hope to take the above forward into my teaching and to further research.

Chapter 4
Learning Mathematical Reasoning in a Collaborative Whole-Class Discussion

My interest in pursuing this study was driven primarily by my personal experience as a teacher as well as being faced with a new curriculum that I was not sure how to implement in my classroom. "Where am I going to use this mathematics?" is a popular question among my learners. This question emanates from learners seeing mathematics as unrelated pockets of knowledge rather than a set of related and useful topics. I have also observed that when learners understand and relate a particular topic to their existing knowledge, this question seldom crops up. I believe that learners' inability to see mathematics as a worthwhile human activity is in part due to the low level of mathematical reasoning and collaboration in classrooms. Learners who learn mathematics through mathematical reasoning may find the mathematics more meaningful. Mathematical reasoning allows learners to form connections between new and existing knowledge (Ball and Bass 2003), and this integration of knowledge may support sense-making on the part of learners and the ability to see mathematical activity as worthwhile. Mathematical reasoning enables the development of conceptual understanding and productive disposition (Kilpatrick et al. 2001), which allows learners to draw on their concepts in other situations and experience mathematics as something they can understand and relate to. Learners who engage in mathematical reasoning may be in a better position to connect school mathematical activity to other activity.

I view collaborative learning as a communicative process whereby two or more parties gain new knowledge as a result of their interaction. Collaborative learning not only refers to an exchange of knowledge between the parties, but the interaction itself serves as a catalyst for the formation of new knowledge by the parties concerned (Mercer 1995). In my class, I think of collaborative learning as a joint venture between learner/s and teacher and among learners themselves. This collaboration is governed by the pursuit of knowledge for the development of learner and teacher. How we reason mathematically or allow our learners to reason mathematically is in part dependent on the nature of collaboration between the parties. The nature of the learning that occurs is a complex interplay between individual and social construction (Hatano 1996; Wood et al. 1992).

This chapter represents a response to the new curriculum developments in South Africa. Motivated by a need to teach in a way that will make mathematics more meaningful to my learners and guided by curriculum change, I decided to explore

K. Brodie, *Teaching Mathematical Reasoning in Secondary School Classrooms*, DOI 10.1007/978-0-387-09742-8_4, © Springer Science+Business Media, LLC 2010

the extent to which this could be achieved in my own teaching. In thinking about how to conduct the study, I posed the following questions to myself:

- What do I understand by mathematical reasoning?
- Why pursue the teaching of mathematical reasoning?
- What is collaborative learning and how does it impact on the teaching and learning of mathematical reasoning?

What Is Mathematical Reasoning?

An important part of all learning, including learning how to reason mathematically, is that new knowledge is always connected to current knowledge, and in fact restructures current knowledge if true learning is to occur (Hatano 1996). So, as we try to develop mathematical reasoning among learners, it is important to see whether and how they make these connections and transform their existing ways of reasoning. As discussed in Chap. 1, mathematical reasoning is intertwined with the other strands of mathematical proficiency (Kilpatrick et al. 2001): conceptual understanding, procedural fluency, strategic competence, and productive disposition. These strands suggest that teaching mathematical reasoning requires far more than merely following a "recipe". If we take seriously the notion of mathematical proficiency, we are faced with an even bigger challenge, the simultaneous development of a range of skills and abilities that is required for learners to be regarded as mathematically proficient.

Kilpatrick et al. (2001) argue that "the strands complement each other but at the same time the reasoning strand, called adaptive reasoning, is the glue that holds everything together" (p. 129). In analysing one learner's developing reasoning in this chapter, I show how mathematical reasoning provides a link with the other strands, particularly conceptual understanding and procedural fluency. I also draw on the notion of mathematical practices (Ball 2003). These include *representational practices, justification, generalization,* and *communication*. These practices are seen as vehicles to achieve the mathematical proficiency discussed above.

The Open University (Open University 1997) suggests that mathematical reasoning unfolds as the learner asks and strives towards answering three important questions while engaged in mathematical activity:

- *What is it that is true?* This question arises as the learner looks to find patterns and regularities that can be rendered as evidence to justify an idea. If enough evidence is found to convince the learner, s/he can formulate a conjecture. This is where we see so many of our learners falter and regard the "evidence as proof" (Chazan 1993). Learners may prematurely draw generalized conclusions based on the measurement of a few examples. For example, learners may conclude that the interior angles of a triangle always add up to 180° after having measured only a few or just one set of interior angles of a triangle.

- *How can I be sure?* This question arises as the learner is confronted with the possibility that the evidence collected may not account for all cases. There now exists the need for some reasoning that would include the evidence in the form of a generalized argument or proof. Without the process of gathering evidence and formulating conjectures, the learner at times regards this proof as merely evidence of another case (Chazan 1993). My learners have often viewed my explanation of a proof of a theorem as an example of how to approach the problems in the exercise and not as an explanation of why the theorem is true.
- *Why is it true?* At times, even a logical explanation that explains the truth of a statement is not enough to convince someone as to why something is true. As De Villiers (1990) points out, the explanatory function of proof or arguments is very different from the verification function. It is likely that learners will need to understand *why* something is true in order to accept it, rather than just verification *that* it is true.

All of the above conceptions of mathematical reasoning, as making convincing and explanatory arguments; as intertwined with the other aspects of mathematical proficiency; as involving a number of important practices; and as restructuring current knowledge and practice, informed this study. However, I still had to answer some other important questions, the next one being why should we teach mathematical reasoning?

Why Teach Mathematical Reasoning?

I argued earlier that I view mathematical reasoning as the vehicle to sense-making of and in mathematical activity. I refer to making sense of the mathematics itself, not necessarily to making links with everyday life. My assumption is that only through making sense of the mathematics can we truly move to sense-making as a worthwhile everyday life activity. The National Curriculum Statement (Department of Education 2003) expresses the vision of a learner who is able to "transfer skills from one context to another" and to "think logically and analytically as well as holistically and laterally" (p. 5). This vision suggests a thorough conceptual understanding of mathematics among learners and the capacity to readily identify situations where their knowledge is of relevance.

Boaler (1997) talks about flexible conceptual knowledge. She worked in two schools that were homogeneous in terms of the socio-economic status and educational background of their learners. The only noticeable difference was the way in which the two schools approached the teaching of mathematics. On the one hand, Amber Hill had a typical textbook approach with lessons consisting of rule-based, procedural activities with much drill and practice. "A typical day of maths in the old apartheid days", was my immediate response. On the other hand, Phoenix Park adopted an open-ended, problem-solving, real-life approach to teaching mathematics, which is what our new curriculum aims at. Boaler's research concluded that learners gained vastly different experiences of mathematics and developed different forms of mathematical knowledge. The majority of learners from Amber Hill were unable to apply their knowledge to new problems and situations. This suggested that they developed knowledge consisting primarily of memorization and applying rules that could only be applied within a school setting. Learners at Phoenix Park, however, developed more flexible knowledge, the kind of knowledge that enabled them to solve new problems they encountered. Boaler's study inspired me to develop a teaching approach closer to that of Phoenix Park. Collaborative learning was the key to developing mathematical reasoning in this approach.

Collaborative Learning and Mathematical Reasoning

The National Curriculum Statement puts forward the following vision for a post-apartheid South Africa: "To heal the divisions of the past and establish a society based on democratic values, social justice and fundamental human rights" (Department of Education 2003, p. 1). This statement acknowledges diversity and the need for equity, promotes the integrity of each individual with the power to affect decisions and suggests that a way to achieve equity is through the promotion of social justice and fundamental human rights. To achieve this, learners need to "work effectively with others as members of a team, group, organization and community" (p. 2). This important notion is picked up later in a focus on mathematics: "mathematics enables learners to work collaboratively in teams and groups to enhance mathematical understanding" (p. 10). Taking these two assertions together, we see that collaborative learning is both an end and a means (Brodie and Pournara 2005). We need to develop skills and dispositions towards collaboration in learners as democratic citizens and also to use collaborative learning to aid mathematics learning.

Developing a social conscience based on *democratic rule, social justice, and human rights* can be obtained within the context of collaborative learning. It would be difficult if not impossible to teach learners to value other people and their opinions, without learners actually learning together from each other. It is the relevance to the learning of mathematics that tends to be more challenging. There is, from South African classrooms, an evidence of teachers using group work without much mathematics learning happening (Brodie and Pournara 2005). I think of collaborative learning as a joint venture between learner/s and teacher as well as among learners themselves. How we reason mathematically or support our learners to reason mathematically is in part dependent on the interdependence between the parties in collaboration.

Mercer (1995) strengthens my ideas about collaboration with the following quotations: "I suggest that we need to recognize that knowledge exists as a social entity and not just as an individual possession" and "the essence of human knowledge is that it is shared" (p. 66). Mercer's ideas resonate with those of Lave and Wenger (1991) who argue that learning occurs in communities of practice, with shared goals and practices. The idea is to create such a community in the classroom, where the teacher takes a leading role in helping learners to develop interactions and practices as a community. Teaching mathematical reasoning demands that learners be able to voice their mathematical thinking, so that mathematical discussion around their assertions can generate an "intellectual ferment" (Chazan and Ball 1999). Learners need to move away from a dependence on the teacher as the only mathematical authority in the class towards a position that Davis (1997) refers to as a "community-established standard: a collective authority" (p. 369). As argued in Chap. 1, this authority comes from the discipline of mathematics. Developing a broader sense of authority requires changes in the way learners and teachers view their roles in the classroom. Teachers need to become what Davis (1997) terms "hermeneutic listeners" (p. 369), which is genuine listening as a participant in the conversation in order to understand what learners are saying. We refer to this kind of listening "with" learners. This is very different from evaluative listening, i.e. listening *for* the right answer, which many teachers do most of the time.

Listening to learners in better ways does not necessarily help teachers to know how to respond to learners' ideas (Heaton 2000). In their article aptly named "Beyond being told not to tell", Chazan and Ball (1999) suggest practical ways in which teachers can act in classroom discussions, without giving the answers, that may focus and give direction to a particular discussion. These include

- Rephrasing learners' comments and helping the class to hear them
- Asking for clarity when they think learners' assertions are not clear and
- Focussing learners' attention on a particular aspect of a discussion

As teachers do this, focussing learners on the norms of participation is important, particularly sociomathematical norms (Yackel and Cobb 1996, see also Chap. 1), where an explanation consists of a mathematical argument, not simply a procedural description or summary; mathematical thinking involves understanding relationships among multiple strategies; errors provide opportunities to reconceptualize a problem, explore contradictions in solutions, or pursue alternative strategies; and collaborative work involves individual accountability and reaching consensus through mathematical argumentation (Kazemi and Stipek 2001).

Summarizing My Perspective

In the above pages, I have made a number of arguments, which informed how I conducted this study and analysed the data. First, I argued that mathematical reasoning is made up of a number of processes. The learner makes observations, tries to provide evidence and explanations, and through connecting these with existing knowledge, restructures this knowledge (Hatano 1996). Proficiency in "procedural fluency" and "conceptual understanding" (Kilpatrick et al. 2001) is needed for such restructuring. Key to enabling restructuring is explaining, communicating, and justifying conjectures and claims, which are features of "adaptive reasoning" as argued by Kilpatrick et al. During this communicative process, we see the learner evaluating and refining new knowledge.

I further argued that learning mathematical reasoning as part of mathematical proficiency (Kilpatrick et al. 2001) is best achieved through collaboration in communities of practice (Lave and Wenger 1991). Such communities are governed by norms of practice, (Yackel and Cobb 1996) and as teachers, we can and should take the lead in developing classroom norms that deeply engage learners (Kazemi and Stipek 2001). Teachers can listen carefully and make a range of moves (Brodie 2004b) which do engage learners' thinking. Taking account of the above, I embarked on a study to see whether what these researchers are claiming is possible in my classroom in South Africa.

My Classroom

My pseudonym in this study is Mr. Daniels; there is a detailed description of my school context in Chap. 2. I refer to some of it briefly here. This study was conducted with one of my Grade 11 classes, consisting of 35 learners with a range of mathematics abilities. The class was situated within a school of 1,600 learners, which is well integrated in terms of historically racial divisions. A teaching staff of 63 puts the teacher–learner ratio at about 1:23. Actual class sizes average 33 learners per class. Although the school is situated in a middle-class suburb, a large number of the learners travel to school from lower income areas. English is the language of instruction at the school and is not the first language of the majority of the learners. My classroom is relatively well resourced with desks and chairs for every learner. The building structure in general is well maintained. Aside from the writing board, I also have an overhead projector and screen at my disposal.

As part of the collaboration in this project, I worked with two colleagues to develop a series of tasks that we hoped would elicit mathematical reasoning in our Grade 11 classes. We drew on a number of resources, including texts that were in the process of being written for the new curriculum. The tasks that we developed have been analysed in Chap. 2. I planned to use the tasks over a week in my Grade 11 class. I structured the work as follows: learners had some time to work on the tasks themselves, then they came together in small groups of three or four learners to discuss their findings, and finally, the groups reported back to the class and we had a whole-class discussion. The lessons were videotaped and I wrote reflections after each lesson, which helped with my analysis.

The Analysis

There were three important issues in my analysis of the data collected:

- The first was how to *select* certain parts of the data to analyse.
- The second was how to see the analysis in *context*.

- The third involved *structuring the writing* to make it easy for the reader to understand my argument even though I can present only some of the data.

As I struggled to select data, I decided to focus on one learner's development in one lesson. It was not possible to do more in the scope of this study and I believed I could achieve more depth of analysis through focussing on one learner. My decision to focus on Winile in particular was because of her visible participation throughout the lessons, which allowed me to plot a developmental sequence of her learning. The analysis therefore focusses on the development of Winile's reasoning through the lesson and how collaboration with me and other learners made this possible. Focussing on Winile's learning enabled me to understand how her learning as an individual both influenced and was influenced by the social interaction in the class.

To isolate a learner from a whole-class discussion in order to analyse and follow her mathematical reasoning is not entirely possible. This is due to the collaborative learning that takes place. In such an analysis, the question arises as to how to know which statements influence each other. One learner's statement may or may not motivate another learner to say something. To link contributions in discussions to each other is a difficult task, and it is important to always remember that there is a variety of influences on learners' development. This means that the context of any utterance needs to be considered very carefully and from a number of perspectives.

The analysis presented in the next section was obtained from thorough analysis of the video and transcript focussing on the claims that Winile made in one lesson, over a time period of about 36 min. It is clearly not feasible to present all of these data here. So, I need to make another selection, which is how to show the reader what I have seen, in much less time and space than it took me to see it.

Winile's Learning

The analysis focusses on Activity 2. The content of the activity was how to think about the changes affected by the horizontal translation of the graph of $y=x^2$ to the graphs of $y=(x-p)^2$ where p was 3 and -4 respectively. Winile's group had just reported back on their findings and Michelle posed a question, asking why the graph for $y=(x+4)^2$ has a turning point of -4. She suggested that the $+4$ inside the bracket contradicted a turning point of -4 and asked Winile to explain this. This served as a catalyst for a fervent discussion, which resulted ultimately in Winile formulating her new conceptual frame for understanding graphs and equations. The analysis follows Winile's learning in five steps: (1) making observations; (2) explaining and justifying claims; (3) connecting her claims to the mathematical representations; (4) restructuring conceptual understanding; and (5) using her new conceptual frame to test other claims. I describe each of these and show how the classroom collaboration supported Winile's shifts.

Making Observations

Winile's journey started as I called her group to share their findings with the rest of the class. Winile became the reluctant spokesperson for the group. In the extract below, Winile hesitantly indicated that the turning points of the graphs $y=x^2$, $y=(x-3)^2$ and $y=(x+4)^2$ differ; the y-values of corresponding points stay the same; the x-values change; and the sizes of the graphs are the same and the equations of the graphs differ.

Winile:	We said they are different on the turning point, and the equation, but the y-axis stays the same, and the size of the graph also stays the same, and (*inaudible*)
Mr Daniels:	Okay so the, what stay the same
Winile:	The, the y-axis.
Mr Daniels:	The y-axis. The y-axis stay the same.
Learners:	*talk over each other, inaudible*
Winile:	The y values stay the same but x-axis changes.
Mr Daniels:	Okay, can we speak one at a time. Let's speak respectfully to one another here. If you've got a question, just raise your hand.
Learner:	(*inaudible*)
Winile:	What
Learner:	(*inaudible*)
Winile:	The y-value stays the same, the x-value (*inaudible*) the turning point (*inaudible*)
Mr Daniels:	Okay, so you say the equation changes, the y-value stays the same,
Winile:	And the turning point,
Mr Daniels:	And the turning point stays the same.
Winile:	No, it changes (*shakes her head and looks at her notes*)
Mr Daniels:	The turning point also changes

We see here that Winile's initial claims were merely observational and she did not see a need for justification. In fact, even to enable her to make a proper report back required a lot of support from me. This support was in the form of keeping other learners quiet, establishing social norms so that Winile could be heard, and also helping her to voice her ideas, and in some cases rephrasing (Chazan and Ball 1999) or revoicing them (O'Connor and Michaels 1996).

Explaining and Justifying Assertions Made

After this, Grant, a member of the same group, came up to comment on the next part of the task. Grant tried to explain that since the x-values changed and the y-values stayed the same as the graphs were shifted left or right, the equations must change. I pressed him to say more specifically how the graph had changed and Grant struggled, looking at his book and searching for an explanation. His attempt follows in the next extract:

Grant:	Sir, uh, the graph's position has moved, so when you, however many positions its moved, you either add it or minus it, onto your equation.
Winile:	Can I just make it simple sir, you substitute the x-value with the variable, we change the equation and then the y, uh, variables never changes (*inaudible*).
Mr Daniels:	Yes, okay, now how d'you mean, just explain what you said, he said that it changes, what did you say? I didn't follow nicely.
Grant:	If the graph, the graph's position has changed, on the x-axis
Mr Daniels:	Right.
Grant:	Therefore, so then you either add onto your equation, its moved how many spaces, or you minus it. Now do you understand?

After Grant's initial contribution, Winile stepped in a little more confidently. Her assertion is still vague; however, it does show a different interpretation from Grant's, which is also vague. This response from Winile suggests that she began to acknowledge a need to explain and justify claims, realizing that Grant's claim needed explanation for her and probably the rest of the class.

Winile's move shows how learning collaboratively feeds into the process of mathematical reasoning. Winile did not see a need to clarify or explain her own claims, yet hearing another learner's claims, which she had been party to, prompted a need to explain. In making her explanation, Winile started to make connections between her observations and the equation. In a sense, she was justifying why the equation must change. She was reasoning at a higher level, brought on by realizing the need to explain Grant's claim to the class. My role in this interaction was to press Grant to explain his claim, which also supported Winile to do so.

Connecting Observations with Mathematical Representations

As learners in the class sought more clarity from Winile with regard to Grant's assertions, Winile realized that she needed to switch representations. She came up to the overhead projector and tried to explain her concept as follows:

> It means that this x uh, here, because when you move three times to your right *(writes $y = x^2 + 3$)*, or you *(writes $y = x^2 - 3$)*, it means that you move to the left, this means when you move to the right three times, that's what we trying to do, that when you move the graph three times, you supposed to add it three times, and when you move it three times to your left then you subtracting three times

These connections between equation and graph are mathematically incorrect. However they do show that she was starting to make conjectures about certain patterns she had observed.

The use of alternate representations by Winile suggests that again she was reasoning at a higher level. Not only was she explaining observations, but she was making connections between her graphical observations and the representations in equations. The need to use a written representation to illustrate the translation of the graph to $y=x^2$ to $y=x^2-3$ (translation of 3 units to the left) and to $y=x^2+3$ (translation of 3 units to the right), served as a catalyst for making these connections. The need to explain to others more effectively once again served as a catalyst for mathematical reasoning. Winile's reasoning was extended to expressing the changes she observed in an alternate representation. She had progressed to not only connecting various aspects of the mathematics but also producing mathematical representations with which to express these connections. Although these representations were mathematically incorrect, they demonstrate her reasoning in relation to the task.

At this point, Winile was interrupted by Michelle who wanted to ask her a question. The interaction involved a few learners and is captured in the following transcript:

Michelle:	Okay, can I ask a question
Mr Daniels:	Okay.
Michelle:	Okay, look on task one right. You said that if it is a positive, you move to the right and if it is a negative, you move to the left. So now, can you please tell me why on your second drawing, where it says y equals x minus three squared (*looks at Winile*) can you see that? Say yes Winile if you understand.
Winile:	Yes, I can see it.
Michelle:	Alright, so now how come in the bracket there's a negative but where the turning point is, is a positive. That's what I would like to know.
Mr Daniels:	Okay, Lorrayne (*interruption by learners*) Carry on Lorrayne
Lorrayne:	Sir, you have a negative three in the bracket and it's a square, when you square something, remember, Sir said when you square it, it becomes positive.
Learner:	If it's a negative
Lorrayne:	Ja
Michelle:	And then if you look at y equals x plus four, why is it that the turning point is a negative.
Learner:	But the equation is positive
Michelle:	And the drawing is positive.
Learner:	I asked that too. (*Some learners laugh*).
Learner:	I'm also asking the same question.

In the above extract, Michelle and Lorrayne co-produced an important question relating the equations to the graph. It was a question that had occurred in a number of groups and so was shared by learners. What is notable in this extract is how the learners worked together and spoke to each other, with almost no intervention from me, except to give Michelle permission to talk and to keep the class quiet so Lorrayne could speak. Brodie (2007b) argues in relation to this episode that when learners share important provocative questions, they are more likely to engage in real conversation.

Winile was silent during the above interaction and continued to be silent as a number of other learners discussed Michelle and Lorrayne's question. Winile did not return to her seat however, but took a seat in front of the class where she listened intently to the ensuing discussion. The discussion continued for some time, until I felt the need to intervene and make an important point as follows:

Mr Daniels:	Okay now, what is a point? A point is made up of what?
Learner:	x- and y-co-ordinates.
Mr Daniels:	x- and y-co-ordinates. Good! So what is the x-co-ordinate there?
Learner:	It's minus four is your x co-ordinate
Mr Daniels:	Good. So what is negative there? The turning point is negative or is it one of the co-ordinates that's negative? Okay, let's hear.

Brodie (2007c, see also Chap. 9) argues that although the above interaction might seemed somewhat constrained, in fact it served an important function in moving the learners' discussion and thinking forward, in that it reminded them to consider both co-ordinates of the turning point, rather than only the x-value. Up until this point, they had been talking about the turning point as −4, which did not help them to see that a point is a relationship between x and y, given by the equation. This interpretation is borne out by Winile's following contributions. The intervention helped to support her to move to the next level of her reasoning trajectory.

Reconstructing Conceptual Understanding

Immediately after my intervention above, Winile emerged from being a silent participant with new ideas to contribute:

Winile:	The positive four is not like the x, um, the x, like, the number, you know the x *(showing x-axis with hand)*, it's not the x, it's another number. For that when you do the equation you get some sense from the answer you get, cause without that p, that minus p, your equation will never make sense.
Learners:	*(murmuring)*
Mr Daniels:	Can I just get back to, That's good, Winile
Learners:	Sshh
Mr Daniels:	Does people want to make clear of what Winile is saying?
Learners:	Yes, *mutter, talk over each other as Winile comes up to OHP*
Winile:	You see, Michelle when you've got this *[writes $y = x^2$]*, you substitute this with a number, isn't it. Like you go, whatever, then it gives you an answer. *[substitutes 3 for x and gets 9]*
Learner:	Yes
Winile:	You see when you got this, plus three *[writes $y = x^2 + 3$]*, you have to substitute this with the, that with like the one, zero, one two, three *[Draws numberline, x axis]*. Your turning point is here. You have to substitute this with this negative one here, plus three. Do you understand? This three *[circles the 3 in $y = x^2 + 3$]* is not, is not part of the, this x, uh, variables. Its the given *(inaudible)* Get it?

In this extract, Winile justified her claims similarly to how she explained Grant's assertions earlier. First, she made a verbal contribution, which was difficult for others to understand. She then came up to the overhead projector and wrote equations to explain her new understanding. As she explained the second time, her explanation is not only clearer to the listener but her explanation has progressed to become more focussed and connected, even though she is still using the incorrect equation.

The clarity of Winile's mathematical reasoning was evident as she explained that the +4 and the −3 in the equations were not the *x*-values of the turning point but as she put it "*some other*" values. She affirmed that the equations represented the relationship between the *x* and *y* variables and that the *x*-values must be substituted into the equations to give the *y*-values. During these assertions, it was evident that Winile was more confident and self-assured that she was on the right track. Winile's reasoning had evolved to a point where she was in a position to evaluate previous claims and adapt them to her understanding. She was now in a position to make the appropriate connections between the value of the turning point and the representational equation. This learning came after a relatively long period of silence from Winile where I can only assume that she was quietly reasoning and adjusting her own understanding as the class discussion involved other learners.

This highlights again the quality of collaborative learning which was present in Winile's reasoning. She was able to modify her assertions by listening to the discussion that prompted her own reasoning. Her explanation to the class facilitated their understanding but also assisted in refining her own understanding of the issue at hand. With this understanding, she confidently answered Michelle and Lorrayne's question.

Testing Other Claims

After this, David indicated disagreement with Winile, arguing that the turning points could be determined by taking out the +4 or the −3 from the bracket, moving them to the other side of the equal sign and changing the signs. Again, she had to justify her ideas, which she did as follows:

> We supposed to get the y, aren't we supposed to get the y, what the y equals. We're not supposed to get what x is equal to, we getting what y is equal to. So we supposed to, supposed to substitute x to get y.

This justification supported Winile to move to yet another level of mathematical reasoning. She emphasized the fact that we use the equation to get the *y*-value by substituting the *x*-value into the equation. In doing this, she tested her own conceptual frame against that of David's and used her understanding to extract the weaknesses in David's argument. Winile did not wait to be invited to give a response to David, but confidently and openly engaged David's assertions. She argued (laughing):

> Okay sir, he's just telling us where to put like, the turning point of the graph, and we want to know why, the y-value is, we want to know what the y-value is and you're telling us the x-value.

Winile was using her conceptual understanding to test and spot the failures in David's argument. This places her in a position to challenge David's assertions. She continued to do this for the rest of the lesson.

The Teacher's Role

The above analysis indicates how important the collaborative learning in the class was to Winile's learning. In particular, the role of the teacher was central to this collaboration. In the above analysis, I indicated a number of roles that I played in supporting learners to talk, in steering the collaboration, in pushing for justification, in remaining silent when I needed to, and finally, in making substantial mathematical contributions when necessary. To further analyse my own role, I came up with three main categories, each of which contains some important teacher moves.

Establishing Discourse

By "establishing discourse," I refer to my actions that attempted to create a climate of interaction, which could support the learners to participate in the discussion and to reason mathematically. One way in which I did this was to create social and socio-mathematical norms in the classroom (Yackel and Cobb 1996). Social norms included speaking one at a time; raising one's hand as an indication that one wants a speaking turn; listening to each other; and building on each other's ideas. Socio-mathematical norms refer to the nature of the mathematical interaction. For example, after Michelle and Lorrayne had asked their question, Candy tried the following response:

Candy:	Sir, couldn't it just be like a basic thing, that if it's on the positive side then your equation is negative and if it's on the negative side then your equation is positive? Can't it just be like that *(laughs)*
Michelle:	I can't accept that
Learners:	Mutter, talk over each other
Mr Daniels:	Okay. Let's … Say that again.
Michelle:	I can't just accept that.
Mr Daniels:	So, I'm not expecting you to accept it.
Michelle:	No, I'm just saying that I can't …
Mr Daniels:	That's good. That's what I'm saying. I'm saying it's good that you don't just accept it

Candy was asking whether we should just accept the fact that the signs were different. Michelle indicated that she could not just accept that, implying that she needed a better justification. I praised her position as valid, indicating that I did not expect nor want her to just accept it and the discussion continued to try to find the justification. This helped to establish the socio-mathematical norm of requiring a justification and may have helped Winile to restructure her understanding to include the need for justification.

The second way in which I established a particular kind of discourse is by modelling how to participate. I listened attentively to try to understand what the speaker was saying and I asked questions if I disagreed with learners' assertions or needed some clarity on their ideas. The following extract occured when I asked Michelle to clarify the question she asked Winile.

Michelle:	And then if you look at y equal x plus four, why is it that the turning point is a negative.
Learner:	But the equation is positive
Michelle:	And the drawing is positive.
Learner:	I asked that too. (*Some learners laugh*).
Learner:	I'm also asking the same question.
Mr Daniels:	What question are you asking?
Michelle:	The question …
Mr Daniels:	Yes.
Michelle:	Look at our drawing where …
Mr Daniels:	Okay. Where's my drawings? (*finds drawings*)
Michelle:	Where it says y equals x plus four on the left hand side.
Mr Daniels:	Right.
Michelle:	Our turning point is a negative four.
Mr Daniels:	Okay
Michelle:	Then Lorrayne that said with the one on the right, where it says y equals x negative three squared, and the turning point is a positive. Because you squaring it, it will become a positive. But what happens with um, the one on the left?
Lorrayne:	The negative one. The equation is positive but the graph is on the negative side.

In the above extract, I model how to listen by asking the learners "what question are you asking", by explicitly showing them that I was looking for my drawings in order to understand their question and by indicating agreement as they spoke and I understood. This is important because many learners have not participated in discussions previously and may not know how to listen and contribute appropriately.

Framing Discussion

By framing discussion, I refer to the actual mathematical content that I used to help the learners make progress. The best example for this is the one quoted above, where I used a sequence of closed and directive questions to remind the learners that a point consists of two co-ordinates. I did not do this because I wanted them to remember that as a fact in and of itself. Rather, it was an important mathematical fact that could help their thinking (see also Brodie 2007c and Chap. 9). By reminding learners that a point consists of a co-ordinate, I focussed their thinking onto the relationship between the *x* and *y*, helping to move the discussion forward.

Lesson Flow or Momentum

With lesson flow, I refer to the movement and progression of discussion. Does the discussion show any progression or is it stagnant on one point, which does not seem to be resolved? Is there any discussion taking place at all? The ability of the teacher to negotiate between speaking turns and free dialogue plays a key role in the lesson flow. The ability to assess when to intervene and when to allow discussion to take its course is an important consideration for a progressive and meaningful lesson flow. For this, the teacher needs to be on the pulse of the discussion, constantly aware of the meanings learners are constructing within the discussion as well as the "social and emotional tone of the discussion" (Chazan and Ball 1999).

Conclusions and Implications

This study set out to analyse the ways in which learners collaboratively engaged in mathematical reasoning and how they learned to reason mathematically through collaboration. The analysis points towards the possibility of such learning. The analysis also provides an argument that this learning was made possible through mathematical processes characterized as follows:

- Making observations
- Connecting observations with various mathematical representations
- Explaining and justifying assertions made
- Reconstructing conceptual understanding
- Using a new conceptual frame to evaluate assertions

My analysis shows how a learner constructed and readjusted her own conceptual understanding of the content, motivated by the collaborative nature of the learning environment. Her learning was not simply learning from her peers but learning with her peers. It could be argued that she might not have moved to a new conceptual frame without the catalyst provided by collaboration characterized by an intellectual ferment (Chazan and Ball 1999) in the classroom discussion. Reflecting on Winile's learning, I can see the important role that collaboration played in her learning.

I have also shown that the teacher is central in collaborative learning. I have shown how I created the conditions of possibility for the collaboration and provided a mathematical voice at certain key moments. From me, there is a strong message to teachers here, which is that this kind of teaching is much harder than traditional teaching. If we are to continue to use the word "facilitate" to describe the teaching we would like to do, we should understand that facilitation requires much more work than we are used to. From this experience, I have seen that lessons in which teachers support mathematical reasoning in their learners through collaborative learning are very time consuming. These lessons are important in that they allow for greater conceptual understanding and reasoning as was shown in the analysis. However, it may be the case that less content is covered, as happened in my class. I suggest that researchers look into developing learning materials and teaching methods that will enable teachers to cover content in a more integrated way, so that more content can be covered, while reasoning is simultaneously developed.

In concluding this chapter, so much is still left unsaid. What is clear to me, however, is that the teaching of mathematical reasoning is achievable through a collaborative learning environment with effective whole-class discussions. A lot of research still needs to be done in establishing sound pedagogy to facilitate this type of teaching. It is my hope that this project will spark a flame in many teachers and researchers to initiate more rigorous research and reflection.

Chapter 5
Classroom Practices for Teaching and Learning Mathematical Reasoning

The previous two chapters focussed on learners, their responses to tasks and one particular learning trajectory, and analysed how the teachers supported the learners through their teaching moves and practices. In this chapter, I shift the emphasis slightly, focusing on the teacher's practices and how these become internalized by the learners. So, while continuing the focus of the book on teacher–learner interaction in the development of mathematical reasoning, I illuminate a slightly different view of the interaction in this chapter.

I do this by analysing a teaching approach that I developed over a number of years, in which thinking and talking are used to promote mathematical reasoning. The approach has been largely influenced by the changes in the South African curriculum over the past 10 years. These changes have been in three areas: what it means to do mathematics (Ball 2003; Kilpatrick et al. 2001), what it means to learn mathematics (Hatano 1996; Lave and Wenger 1991; Vygotsky 1978), and what it means to teach mathematics (Chazan and Ball 1999; Lampert 2001). Informed by these shifts, I have developed a new approach, one which might be called learner-centred, where learner-centred means encouraging learner participation in ways that allow learners to reason mathematically, to make sense of mathematics, to transform their mathematical ideas, and to own their mathematical thinking (Brodie 2007a; Brodie et al. 2002). This is different from older curriculum approaches in which teachers introduced the subject matter, gave exercises to learners, and corrected them, without hearing much about learners' underlying thinking.

My approach aims to support learners' thinking and talking in mathematics and to develop their mathematical reasoning. By mathematical reasoning, I mean "establishing some truth about a particular aspect of mathematics, finding some evidence or justification for the truth that one has assumed, and knowing why you are correct" (Open University 1997). As discussed in Chap. 1, this view of mathematical reasoning is a broad one, it includes the notion of proof and proving, but is not restricted to them. Rather, I try to support learners to make reasoned and justified statements of their mathematical ideas.

Communication in the classroom, between learners and the teacher and among learners themselves, forms an important component in teaching learners to think and reason mathematically. Thinking mathematically is something that every

K. Brodie, *Teaching Mathematical Reasoning in Secondary School Classrooms*,
DOI 10.1007/978-0-387-09742-8_5, © Springer Science+Business Media, LLC 2010

human being does all the time (Mason et al. 1982). What is important in the classroom is how this thinking is externalized to make one's ideas understood by other people. Lampert (2001) argues for learners to evaluate their own thinking in three ways: by privately reflecting on what they are doing; by talking about it in the local community, which in this case would be in their groups; and by presenting their ideas to the class for public discussion under the guidance of the teacher. It is in this way that I hope to encourage learners to share their ideas and respect each other's opinion in their discussions.

Finally, an important part of reasoning mathematically is that the learner comes to own her or his ideas. A justified argument makes sense because it is justified. My approach encourages learners to refrain from viewing me as someone who has solutions to all the problems and who is the authority on whether something is correct or not. Rather, they should consider me as a person who is there to help them in making their own sense of mathematics.

This chapter focusses on my attempts to achieve the above. In doing this, I present an analysis of my teaching practices and the learning practices that are encouraged in my classroom.

Classroom Practices

I use Schifter's definition of teaching practices as being skilful, patterned regularities that occur in teachers' classrooms (Schifter 2001). These involve particular approaches that teachers employ in their classrooms consistently and which create contexts for developing meaning in mathematics. Practices are always social, intellectual and practical and are directed towards a desired end or goal (Brodie 2008).

An important distinction can be made between teaching practices and mathematical practices (Ball 2003; Cobb 2000). Teaching practices are more general and occur in all classrooms, for example asking questions, writing on the board and asking for learner contributions. Mathematical practices are specific to mathematics classrooms, for example, explaining, generalizing, and justifying mathematical ideas. In describing "classroom mathematical practices", Cobb (2000) focusses on mathematical interpretations and reasoning. Mathematical practices involve the normative or taken-as-shared mathematical content in arguments that arise in mathematics classrooms, are established by a classroom community, and "can be seen to constitute the immediate, local situation of the students' development" (p. 73). I will talk more on the notion of classroom community below.

The Productive Pedagogies Research Group (Hayes et al. 2006) looked at a number of classroom practices that contribute to more equitable student outcomes for all students. Some of these practices are developing higher order thinking and depth of knowledge among learners; employing extended conversations and metalanguage in classrooms; creating connectedness and integration among topics; and encouraging learner self-regulation. These resonate with "equitable teaching practices" described by Boaler and her colleagues, which include asking conceptual questions; keeping the level of mathematical challenge high; enabling broader participation of learners in class; and supporting learners' accountability to each other and to the mathematics (Boaler 2002, 2004; Brodie et al. 2004). Drawing on these practices as background, I analyse my own classroom to see which practices I employed to encourage learner participation, reasoning, and accountability.

Learning Mathematical Reasoning

As discussed above, mathematical reasoning is about the conviction that comes with knowing that you have a justified argument, which you can communicate to others. This notion is strongly informed by Kilpatrick et al.'s (2001) five strands of mathematical proficiency, which are discussed in detail in other chapters of this book. Here, I focus on their notion of adaptive reasoning, which they argue, holds the other four strands together. They argue that adaptive reasoning refers to the capacity to think logically and includes knowledge of how to justify conclusions. It is important that learners know and understand that answers are right because they make sense and come from valid reasoning, rather than merely accepting what the teacher and textbook tell them. Learning to reason mathematically involves a number of processes. These are the learner's individual thinking and sense-making; teacher–learner interaction around reasoning; and the classroom as a community of practice developing the mathematical practices of justification, generalization, and communication.

As individuals, learners construct their own meanings of mathematical ideas, talks and symbols (Hatano 1996). Although this always happens in a social context, it is still important for teachers to focus on particular learners' constructions and reasoning. Individual learners bring their current knowledge into the classroom, and hopefully through the process of interacting with others, will transform, shift, or reconstruct this knowledge (Hatano 1996). Hatano argues that the fact that misconceptions exist shows that learners do construct their own knowledge because they are not often taught misconceptions. Errors and misconceptions are signs that learners are involved in their learning and that their thinking processes are engaged. Accepting errors and misconceptions as a normal part of the teaching and learning process means that further explanations can be encouraged from learners in order to understand why they made those errors and misconceptions. This is one way that further thinking and reasoning can be supported among learners.

Many teachers and theorists have interpreted constructivism to mean that learners work on their own, without the teacher. This in itself is a serious misconception. The importance of teacher–learner interaction comes to us from Vygotsky (1978), who argues that learning arises out of two minds in interaction, in this context the minds of teacher and learner. Vygotsky (1978) argues that learning takes place on two planes, on the inter-psychological (interaction between people) as well as on the intra-psychological (interaction within the mind of the individual). The intra-psychological is internalized through the inter-psychological. What is therefore internal in the "higher mental functions" was at some stage external, between people (Vygotsky 1978, p. 80). Internalization of classroom practices by learners is therefore an important support for and indication of learning.

Finally, although learners come into a classroom as individuals, they immediately become part of a community that exists in the classroom, what Lave and Wenger (1991) refer to as a community of practice. In this view, learning and teaching are seen as participation in socially situated practices (Lave 1996). As members of a community of practice, learners learn to participate in the classroom practices, through Legitimate Peripheral Participation (Lave and Wenger 1991). They learn practices from the teacher as well from each other. Drawing on both Vygotsky and Lave and Wenger, this study explores how learners, as members of a community of practice, internalize aspects of their teacher's practices in developing their mathematical reasoning.

Teaching Mathematical Reasoning: Questioning and Listening

When learners reason mathematically, they explain, they generalize, they justify, and they communicate mathematics. "Students need to be able to justify and explain their ideas in order to make their reasoning clear, hone their reasoning skills, and improve their conceptual understanding" (Kilpatrick et al. 2001, p. 130). Learning mathematical reasoning is, of course, a process. It is a process that needs the guidance of the teacher and the participation of the whole classroom community. The teacher's guidance involves practices that teachers employ in the classroom to help learners make sense of mathematics.

In looking at my teaching practices, I chose to focus on two main categories: teacher questioning and listening. Questioning plays an important role in mathematical reasoning, and teachers can ask questions that support or inhibit learners' mathematical reasoning (Boaler and Brodie 2004; Watson and Mason 1998). Teacher listening complements teacher questioning, in that when one asks a question, one ought to listen to how that question is answered. Listening carefully to how learners respond to questions helps teachers to know how to take their ideas forward in supporting them to think and reason mathematically.

Questions are normally asked in many mathematics lessons, in the form of tasks that learners have to work on, or as exercises that learners have to complete, or questions asked by the teacher to assess learners' understandings. Research has shown that different kinds of questions influence mathematics learning and reasoning in different ways. Watson and Mason (1998, p. 3) believe that questions such as "How did you …?, Why does …?, and What if …?" are typical questions that can support learners to focus their thinking on the structures and processes of mathematics. This is in contrast to questions that only focus on recollection of facts, where the teacher usually expects particular answers (Boaler and Brodie 2004). Many authors refer to teacher questions as being closed and thus by having a single, straightforward answer, their main aim becomes testing the learners, rather than encouraging them to think (Nystrand and Gamoran 1991).

If teacher questions are more thought provoking, they can support learners to present a variety of responses. In this way, learners may be encouraged to think and reason mathematically and to evaluate each other's responses. Learners usually learn how their teacher asks questions and come to expect particular kinds of questions from their teacher. If learners come to expect more complex questions, they are likely to expect to have to provide more complex answers, which require reasoning. However, finding the right questions to ask is not always easy. Heaton (2000) shows how many of her questions were too open, they did not support learners to engage with the task. She struggled to find a way to ask the appropriate questions. Kazemi and Stipek (2001) use the notion of "high press" and "low press" to distinguish between questions and prompts that teachers use to push learners into verifying their answers. They argue, "high press questions encourage learners to include mathematical arguments in their explanations, while low press questions encourage procedural descriptions only" (p. 78).

Asking questions goes hand in hand with how learners' responses are heard and responded to, by both the teacher as well as other learners in the classroom. For this reason, listening is also an important tool in an environment that supports mathematical reasoning and thinking. Davis (1997) makes a distinction between listening *for* and listening *to*. Teachers are often constrained by the fact that they listen *for* something in particular, rather than listening *to* the speaker. Listening *for* something goes with not being interested in what the other person is saying. Teachers often ask questions that address particular aspects or points that we are looking for, and when a learner produces an unexpected contribution, we usually do not entertain that response, but continue to look for a response that would satisfy us. Listening *to* a learner suggests trying to understand the sense that the learner is making of the mathematics and taking that as the starting point for further discussion. Davis calls listening *for* something, evaluative listening, and listening *to* someone, interpretive listening. He also has a notion of hermeneutic listening, which we have called listening *with* the learner, where the teacher listens as a co-participant in a conversation with the learners. Listening *with* learners can help the teacher to listen more carefully, interpret the learners' ideas more appropriately, and interrogate their responses.

As a teacher really listens to learners, s/he will find that the errors that learners make are quite sensible and come from underlying misconceptions. Brodie (2005) argues that errors are often "remarkably reasonable when viewed from the perspective of how the learner might be thinking" (p. 37). Schifter (2001) recommends that teachers try to follow learners' lines of reasoning, even when the sense they are making is not obvious. Errors and misconceptions are an indication of learners' thought processes and can be viewed by teachers as uncovering important mathematical questions for the class to consider and discuss. Care must be taken though that other learners do not discourage learners who produce errors, but rather help them constructively. The main aim is to sharpen learners' evaluation skills and abilities to help their peers and themselves, and in doing so, make learning more meaningful. Learners' misconceptions can help them develop into better mathematical thinkers, if teachers ask learners to explain their thinking when they produce these misconceptions (Brodie 2005). How to deal with errors and misconceptions forms part of the practices that teachers can develop to support learners' mathematical reasoning.

My Classroom

For the purposes of linking this chapter with the rest of the book, it is important to note that my pseudonym in the study is Mr Mogale. This study was conducted in one of my Grade 11 classes, in a functional township school west of Johannesburg, with very basic facilities (see Chap. 2 for more detail). All the teachers and learners in the school were "black African" South Africans (see Chap. 2). English is not the main language of any of us, but all teaching and learning of mathematics occur in English. There were 1,700 learners in the school at the time of the study and 46 teachers, giving a learner–teacher ratio of 37:1.

This was a reflective study on my own practice and was conducted in one of my Grade 11 classes, with 43 learners in the class. This was an accelerated class, where learners were taught more quickly than usual due to their strong achievement in mathematics and were eventually introduced to the Grade 12 syllabus while they were still in Grade 11. Most of the learners in the class were very strong in mathematics. I had taught this class (except for six learners) for one and half years, since they were in Grade 10.

I worked together with the other two Grade 11 teachers involved in this research project to choose a suitable task that would elicit learners' mathematical reasoning. We chose a task that consisted of four activities to be done over a week. The task was based on quadratic functions, but was different from the problems that learners had dealt with before, since it involved exploring translations of the graph of $y=x^2$. The task is discussed in more detail in Chap. 2. Learners worked in groups of three or four, which are easy to manage, and each group nominated a spokesperson to present their ideas to the class. In making these presentations, they had to explain to the class how they arrived at their solutions and why they thought their solutions were correct. In the process, they gave other learners the opportunity to ask them questions.

The role that I played during the lesson was that of a facilitator. I walked around the groups to check on how learners were discussing the activities and sharing ideas. I would ask questions that encouraged them to reason mathematically. During group presentations to the class, I had to make sure that discipline was maintained. Any learner who wanted to ask a question or comment should first raise his/her hand to be recognized. I tried as far as it was possible to stand back and give learners the opportunity to communicate their ideas, and allow the discussions to flow. I would only come in when it was necessary for me to insert my own voice (Chazan and Ball 1999) in order to keep the mathematical discussion and reasoning at reasonable levels. Some of my questions and interventions were planned and some happened in an impromptu way, depending on learner responses.

Three lessons of 70 min each were devoted to these tasks. All these lessons were videotaped. In order to analyse my practice, I watched the videotapes very carefully, looking at how I questioned and listened to my learners. I also looked for some other aspects of my teaching that I was trying to achieve, for example challenging learners for justification and redirecting learners' input to the whole class. While noticing these, I became aware of some practices that I did not know about in advance, for example, adding my own voice or adding a learner's voice. So, my final set of categories included aspects of my teaching that I had anticipated and those that I had not.

As I did the analysis, I realized that the practices of questioning and listening could be further distinguished into teacher "moves" (Brodie 2004b, see also Chap. 9). Practices here are viewed as a bigger set of which teacher moves form a subset. Teaching practices enable moves to surface in the classroom, and it is through moves that teachers and learners act in a manner that helps in the development of the practices. I also noticed learner moves and practices and saw how some of these related to my moves and practices. I discuss teacher and learner moves and practices in the following sections.

Teacher Moves and Practices

As discussed earlier, the two key teaching practices that I used to enable mathematical reasoning were questioning and listening and these are seen through particular teacher moves. In this section, I describe the teacher moves that I saw in my practices, give examples, and show how questioning and listening both support and are supported by these moves. Table 5.1 shows the teacher moves that I identified in my lessons. They can be categorized into two inter-related categories: enabling learner participation and communication and focussing on learners' mathematical reasoning.

In the following transcript, I show how I make some of these moves and how they relate to the practices of questioning and listening. This extract from the lesson occurred during a report back from Mpolokeng's group in response to Activity 1, where learners were asked to explain what they observed when the graph of $y = x^2$ was shifted 3 units to the right and 4 units to the left. Earlier, Mpolokeng had explained that in the case of the graph shifting 4 units to the left, the values of x would all be negative and the shifted graph would not cut the y-axis. The class had

Table 5.1 Teacher moves

Enabling participation and communication	Focusing on mathematical reasoning
Redirecting to the whole class	Using learner contributions to move forward
Adding a learner's voice	Adding my own voice
Stirring productive argument	Challenge for justification
Supporting and sustaining "intellectual ferment"	Representing mathematical knowledge
Maintaining appropriate emotional tone	Providing resources for thinking

challenged her and she had come to agree with them that the graph would cut the
y-axis. She then repeated her observation below (somewhat defensively), which led
to another discussion.[1]

1	Mpolokeng:	What? I have said that the x-values are negative, haven't I? Yes, I have said that the x-values are negative, and then when we extend the graph, it will cut the y-axis
2	Takalani:	(Nods his head)
3	Mr Mogale:	(Learners raise their hands) Let's listen to Tebello first.
4	Tebello:	If the graph can cut the y-axis (learners laugh)
5	Mr Mogale:	Talk, what are you laughing at?
6	Tebello:	if, if the graph can cut the y-axis, the part on the right of the x axis, is it not positive?
7	Mpolokeng:	Ee [yes] e [its] positive.
8	Mr Mogale:	It's a good question because she is saying that is going to be on the negative only, right?
9	Learners:	Yes.
10	Mr Mogale:	She says we have only negative values.
11	Learners:	Ee. [yes.]
12	Mr Mogale:	But, we said that this graph can be extended, so what are we saying? We don't only have to question, we must also be in a position to assist a kere? [right?]
13	Learners:	Yes.
14	Mr Mogale:	We can question, we can comment, we can advise the group, (points to learner) Ee [yes]
15	Gordon:	Nna [me], I was saying according to the papers you gave us, e tlo nna fela [it will only be] negative, e tlo nna fela [it will only be] negative.
16	Mr Mogale:	But, we are also given arrows and those arrows should mean something to you.
17	Gordon:	We only concentrate on the left hand side
18	Mr Mogale:	When you go out of the school and you go to the road crossing of ext fourteen and Randfontein road, there's a board that shows you Randfontein that direction, Jo'burg there. It has an arrow, which shows you, right, what does it mean, we stop there, Randfontein ends there?
19	Gordon:	It's not the same as the graph
20	Mr Mogale:	It's not the same as the graph? But the arrow, what does the arrow tell you?
21	Gordon:	(Inaudible) (learners laugh)
22	Mr Mogale:	I am asking you about the arrow gore [that] what does it mean to you?
23	Gordon:	yes, it continues
24	Mr Mogale:	It means that it continues. So, what are you saying? We will only get negative values?
25	Gordon:	I was only referring to this graph, on the table
26	Mr Mogale:	Ee. [yes.]
27	Gordon:	You see the table that they have given us, you see, Mpolokeng never looked at the graph, she's looking at the table

[1] Much of each transcript in this chapter was translated from Setswana. Wherever Setswana words
and English translations remain in the text it is because the original utterance was in English with
some code-switching.

The above extract illuminates a number of moves in the table above, as well as my questioning and listening practices. The first thing to notice is that Takalani is the learner who had asked Mpolokeng to repeat her observation. He nodded his head to indicate that he accepted her new formulation. However, a number of other learners raised their hands, indicating that they wanted to say something. I indicated that Tebello should talk, but as he started talking, other learners started laughing. This was not usual in this class, so I reacted with a question: what are you laughing at? This was enough to suggest to this class that it was a ground rule (Edwards and Mercer 1987) that we allow each other to talk without laughing. This helped to *restore an appropriate emotional tone* to the discussion. Tebello continued to make his point, which was that if Mpolokeng was then claiming that the graph did cut the *y*-axis, then she could not claim that all the *x*-values would be negative, since those on the right hand side of the *y*-axis would be positive. Mpolokeng immediately agreed with him.

In all of the above, and previously, I had been listening *to* Mpolokeng and the learners who responded to her. I did not correct her mistakes, but I allowed the conversation to flow and learners to interact. At this point however, I worried that Mpolokeng was very quick to agree with Tebello. This could have been an instance of unproductive agreement (Chazan and Ball 1999). It was therefore time for me to intervene more forcefully, which I did in turns 8, 10, and 12. I repeated Tebello's point and praised it as a good question. At this point, I was *adding my own voice* mathematically, indicating my agreement that Tebello's question was an important one (note, I was not giving the answer). In doing this, I was also *adding to a learner's voice*, or revoicing his contribution (O'Connor and Michaels 1996). I was also trying to *sustain "intellectual ferment"* (Chazan and Ball 1999), so that the discussion would not end in an unproductive agreement. In turns 12 and 14, I made an additional point that learners need not only challenge Mpolokeng, but could also help her to think through her ideas (something I have always stressed in this class). In doing this, I was again trying to create a *positive emotional tone*, as well as *support a productive argument*.

My interventions above allowed Gordon to talk, and he tried to suggest a reason why Mpolokeng's group had made a contradictory argument. He suggested (turn 15) that on the task handouts, the *x*-values that had been chosen were only negative and that Mpolokeng had been working from those, which focussed her attention on the left hand side of the graph (turn 17). In turn 18, I made *a challenge for justification* and also *pointed to a mathematical representation*, by focussing Gordon's attention on the arrows of the graph, which suggested that the graph continued to cut the *y*-axis, even if it was not shown on the picture in the handouts. In turn 18, my method of interaction shifted from listening *to*, to listening *for*. I did not consider an interpretation of Gordon's comment in turn 17 as being that perhaps Mpolokeng's group knew about the arrows but chose to ignore them. I also did not consider that perhaps Gordon's group had done the same thing, which he was then justifying. Instead, I tried to find an everyday analogy, which might help Gordon and others to understand the function of arrows on the graph, by using the idea of arrows on a road signboard. Gordon's response to the analogy was that it was not the same as the graph, which I agreed with, but suggested that he could still learn something from how the arrows function, that it shows continuation, both on the road and on the graph. In this case, I was *providing a resource for thinking*, an analogy from everyday life.

Although this resource may have been helpful for some learners, it was limited in two ways. First, as with all everyday analogies, it could model only some of the aspects of the mathematical situation. All analogies must be limited in some way, and it is important for teachers to understand how they are limited. Second, and more importantly for this chapter, I presented this analogy in an attempt to teach Gordon something that he already knew. In not listening *to* him, I did not see that he probably did understand this, but was trying to present a reasoned argument for why Mpolokeng had made her claims. In fact, he was finding her errors; as he said in turn 27, she had probably not even looked at the graph. If these had indeed been his own group's errors, as I suggested above, then he was clearly learning something that I did not realize. In fact, I did not even see what Gordon was doing and the strength of his contributions, until I did this analysis.

In the above analysis, I have shown how I used a range of teacher moves, with varying degrees of effectiveness. These moves show that at the beginning of the episode, my questioning managed to support some productive discussion and challenge of ideas. I was able to listen *to* the learners. However, in the second part of the episode, I became more directive, did not listen to the learners and tried to focus the learners on a particular representation of mathematical ideas. This analysis shows exactly how difficult it is for a teacher to maintain the practices of appropriate questioning and listening, even when s/he starts out in that way. The above analysis also shows how a learner, Gordon, was able to hold his ground, and in effect challenge me on some of my ideas, particularly the analogy. This was a common practice among the learners in my class. In the next section, I illustrate some of their moves and practices.

Learner Moves and Practices

A second part of my analysis was to identify a number of learner moves, as well as their questioning and listening practices (Table 5.2). The moves that I identified were all similar to, and a subset of, mine. These were

Table 5.2 Learner moves

Enabling participation and communication	Focussing on mathematical reasoning
Stirring productive argument	Using learner contributions to move forward
Supporting and sustaining "intellectual ferment"	Challenge for justification (both the teacher and other learners)

The following extract shows three learners talking to each other, with very little input from me. They were discussing a question from Activity 2: what is similar in the graphs of $y=x^2$; $y=(x+4)^2$ and $y=(x-3)^2$.

1	Mamokete:	Oh, they are similar in, why I am saying they are similar in the y-values, we don't have the value of q there, it shows that if it is not there, it is zero that value of q, that is why they are the same throughout.
2	Mr Mogale:	Questions, comments, Mapula
3	Mapula:	Which y-value, where is the y-value? For what? y-value of which point?
4	Mamokete:	For the turning point.
5	Mapula:	Only?
6	Mamokete:	What do you mean?
7	Learners:	(*laugh*)
8	Mapula:	It means only they are similar, You say they are similar in y-values, don't you?
9	Mamokete:	Yes
10	Mapula:	So, I am asking that, you are implying that it's y-value is zero?
11	Mamokete:	Yes
12	Mapula:	For the turning point?
13	Mamokete:	Yes
14	Mapula:	Oh, what about there, our y-value is not the same
15	Mamokete:	The other y-value?
16	Mapula:	For the other points (*inaudible*) on this graph, that lie on the graph, the one on the graph, are they not the same? (*she is pointing in the sheet*)
17	Mamokete:	They are the same, these graphs move to left and right, so there is no way that they cannot be the same
18	Mapula:	Oh.
19	Mr Mogale:	Do you understand her question?
20	Mamokete:	Yes
21	Mr Mogale:	What is she saying?
22	Mamokete:	She says I am implying that at the other points, beside the turning point, the y-value is not the same, and I said they are the same (*Aganang raises hand*)
23	Mr Mogale:	Mm
24	Aganang:	But that other time you said that since there is no q, it means then that the y-value is zero, but on the other points (*she is pointing on the board*)
25	Mamokete:	(*Interrupts Aganang*) We are talking about the turning point, I am talking about the turning point

In her initial questions, Mapula was pushing Mamokete to be specific about which points she was claiming had the same y-values. Initially, Mamokete was referring to the turning points only; she spoke about the q value being zero in all three graphs and in line 4, she explicitly said she was talking about the turning points. However, through Mapula's *challenges for justification*, particularly line 16, Mamokete seemed to shift her view, saying that since the graphs shifted horizontally, all the y-values (presumably of corresponding points) would stay the same. This indicates a shift in her thinking, made through the conversation. So, it seems that Mapula was able to *use Mamokete's contribution to move the discussion forward*.

However, when Mamokete was further challenged by Aganang, that she was contradicting an earlier point, it seemed that she might shift back to an earlier position. Such shifting of positions is characteristic of genuine dialogue and suggests that the learner is thinking through her ideas far more than a learner who tries to provide an answer that she thinks the teacher wants to hear. Through this interaction, the girls were exploring the nature of the graphs and their relationships to the equations. They were also *stirring productive argument* and *supporting and sustaining intellectual ferment*. They took seriously the roles of asking and answering questions to clarify each other's thinking and were taking up each other's ideas.

I took only three turns in this exchange (lines 2, 19, and 23), which is very unusual in mathematics classrooms, although there are quite a few places in my lessons where this happens. My three turns did not make any substantial mathematical contributions, but rather were directed at getting learners to talk and listen to each other. The first opened the floor for contributions, while the second and the third intervened to *add to the learner's voice*. In line 19, I asked Mamokete whether she understood Mapula's question, indicating its importance and in line 23, I supported her to repeat the point that she had learned through the conversation. This, then, allowed Aganang to come in, suggesting a contradiction with Mamokete's earlier position.

The above analysis shows that the learners had internalized and could use some of my moves and practices. They listened to each other, challenged each other, and supported strong argument and justification of ideas, with the support from me. This is in line with Vygotsky's (1978) ideas that learners can and will internalize their teacher's ways of talking and interacting. It also supports a notion of community of practice (Lave and Wenger 1991), showing that the learners can interact with each other in ways that support the development of mathematical practices.

Conclusions and Implications

In this study, I have identified a range of practices and moves that I made as I shifted my teaching in relation to the new curriculum. I have shown that by engaging regularly in these practices and moves, learners also internalize them and begin to use them. Thus, they become classroom practices, engaged in by learners and the teacher as a community. I have also shown how at times I was unsuccessful in shifting my practices and became more directive. This was evident to me only on analysis of my teaching and suggests that this kind of action research can suggest how to improve one's practices. Shifting one's teaching is a process that needs refining over time (see also Slonimsky and Brodie 2006). Through discussions with other people involved in the project, my teaching went through some positive changes, directed towards allowing learners to think and reason mathematically. I am still in the process of developing my approach to maximize learner involvement in the teaching and learning of mathematics. In doing so, I recognize that there will always be aspects of previous practices that remain in my new practices, not everything that we used to do was problematic and it is not possible to transform one's teaching into something completely different (Brodie 2007a, 2008).

This study illuminates some important general aspects with regard to teaching and learning mathematics. Teaching a mathematical topic does not necessarily require of teachers to first give an introduction and show learners how to do a particular task. Rather, giving learners a choice to do the task first and then discussing it is a useful approach, as my study shows. My study also shows that communication plays a very important role in supporting learners to explore mathematical ideas. An environment that allows learners to communicate about mathematics can be created in order to give learners the opportunity to think and reason about what they are doing, thus making sense of mathematics.

It is therefore necessary for us as teachers to give every learner the chance to present their ideas, as well as allow them to convince us of their thinking. We will not perfect this process, in one go, but with time may develop it as part of our practices, of questioning and listening to learners. This study has helped to convince me, and I hope will convince you that supporting learners to think and talk about mathematics will go a long way in helping them to make sense of mathematics.

Chapter 6
Teaching Mathematical Reasoning with the Five Strands

Kilpatrick et al. (2001, p. 116) describe a composite, comprehensive view of successful mathematics learning and what mathematical proficiency means, in terms of five interwoven and interdependent strands. The strands are conceptual understanding (CU), which entails comprehension of mathematical concepts, operations, and relations; procedural fluency (PF), involving skill in carrying out procedures flexibly, accurately, efficiently, and appropriately; strategic competence (SC), which is the ability to formulate, represent, and solve mathematical problems; adaptive reasoning (AR), which is the capacity for logical thought, reflection, explanation, and justification; and productive disposition (PD), a habitual inclination to see mathematics as sensible, useful, and worthwhile, coupled with a belief in diligence and one's own ability to come to know mathematics.

Kilpatrick et al. (2001) argue that mathematical proficiency cannot be achieved by focusing on one or two of the strands but that development across all five strands raises the standard of mathematical proficiency, because the strands interact and reinforce each other. The authors suggest, "students who have opportunities to develop all strands of proficiency are more likely to become truly competent at each" (Kilpatrick et al. 2001, p. 144). Therefore, teachers need to structure classroom activities so that all five strands are emphasised and synchronised.

The new national curriculum for South Africa resonates with this notion of proficiency and mentions elements of each of the strands in its outcomes for mathematics (Department of Education 2003). Since it is written in the language of outcomes, it emphasizes processes such as efficient calculation (procedural fluency), creative problem solving in both mathematical and real-world contexts (strategic competence), using mathematics to understand the world (strategic competence), and logical reasoning and justification (adaptive reasoning). It also emphasizes the importance of deeper understanding of mathematical ideas (conceptual understanding) and beliefs that we can make sense of mathematics (productive disposition). A curriculum that emphasizes the five strands is an important first step in teaching mathematical reasoning. However, the intended curriculum often does not become the enacted curriculum for a variety of reasons. Sometimes, teachers don't fully understand the new curriculum (Chisholm et al. 2000), or conditions in classrooms make it difficult for teachers and learners to enact the policy.

K. Brodie, *Teaching Mathematical Reasoning in Secondary School Classrooms*,
DOI 10.1007/978-0-387-09742-8_6, © Springer Science+Business Media, LLC 2010

My focus in this chapter is on the extent to which I could include the five strands in a series of lessons with my grade 10 class.

A Social-Constructivist Framework

My theoretical assumptions are that learners actively construct or create their own knowledge, and that their ability to do so is enhanced when they work together and communicate their understandings. Hatano (1996) argues that although knowledge can be transmitted from teacher to learner, for the learner to make it his or her own always requires interpretation. He argues that "active humans almost always try to interpret and enrich what is transmitted, in other words, to supplement it by construction" (Hatano 1996, p. 200). This interpretation and construction should restructure the learner's existing knowledge into better-organized knowledge. So learning is never only about adding new information, but about integrating this information with the learner's existing knowledge so that it becomes reorganized and more powerful.

Vygotsky (1978) argued that a learner's restructuring comes about not just as a result of his or her own internal processes and abilities, but is also indicative of the communication between teacher and learner. His theory investigates the construction of knowledge as a joint achievement between teachers and learners. Drawing on Vygotsky, Mercer (1995) recognizes specifically that "people construct knowledge together" (p. 67) and that social interaction strongly constrains learners' constructions. Bruner uses the concept of "scaffolding" to highlight how one person can be closely involved in someone else's learning, through supporting his or her nascent ideas. Scaffolding "represents both teacher and learner as active participants in the construction of knowledge" (Mercer 1995, p. 74).

For this research, a social-constructivist framework required a careful selection of tasks and planning of their implementation in the classroom. It was essential for the learners to communicate their thought processes (Mercer 1995) through, firstly, speaking, then in writing and then in looking at each other's work and in evaluating it. To afford every learner the opportunity to communicate, they needed to work together in small groups of three or four, so that they could discuss and reason informally about their various ideas. My function as the teacher was to assist in this joint venture by monitoring group progress and providing appropriate scaffolding. I needed to be aware of maintaining the integrity and complexity of the tasks, so that the mathematical reasoning processes in which the learners were engaging were not diluted by my giving too much, or not enough, help (Stein et al. 2000). Furthermore, I needed to ensure that adequate time was allocated for both small group and whole class discussions. The purpose of the whole class, teacher-led discussion was to share and develop further the achievements and difficulties in all the groups so that all the learners could achieve a high level of knowledge construction and hence, mathematical proficiency.

Mathematical Practices and Proficiency

Kilpatrick et al. (2001) argue that traditionally schools in the United States have emphasized procedural understanding at the expense of the other strands. This is also the case in South African schools (Brodie 2004a; Taylor and Vinjevold 1999), although some teachers do try to teach for conceptual understanding as well. The over-emphasis on procedural fluency has led to the mathematics reform movement in the United States and influenced the new curriculum in South Africa. The mathematics reform movement emphasizes that students should develop "deep and interconnected understandings of mathematical concepts, procedures and principles, not simply an ability to memorize formulas and apply procedures" (Stein et al. 1996, p. 456). Stein et al. also say that mathematical understanding occurs when students "engage in the process of mathematical thinking" and "doing what makers and users of mathematics do: framing and solving problems, looking for patterns, making conjectures, examining constraints, making inferences from data, abstracting, inventing, explaining, justifying, challenging, and so on" (Stein et al. 1996, p. 456).

Ball (2003) refers to what people who learn and use mathematics successfully do as mathematical practices, for example, using symbolic notation effectively, making generalizations, and justifying claims. He says that knowledge of topics, concepts, and procedures is central to knowing mathematics but is not sufficient to use mathematics effectively. Using "mathematics involves doing a series of skillful things, depending on the problem" (p. 30). These practices are crucial to mathematical proficiency, and they further suggest that if we could understand these practices, and how they are learned, teachers would be able to facilitate significant improvement in learners' achievement in mathematics. This could allow us to address unequal acquisition of mathematical proficiency in school. In terms of the five strands, the two that are most closely related to mathematical practices are strategic competence and adaptive reasoning.

Strategic competence refers to the ability to formulate, represent, and solve mathematical problems. Strategic competence does not develop in a vacuum; it requires a rich background of knowledge and intuition (Stewart, in foreword to Polya 1994/1990). Schoenfeld concurs when he says that any "mathematical problem-solving performance is built on a foundation of basic mathematical knowledge" (Schoenfeld 1985, p. 12). In terms of the strands, this would be conceptual understanding and procedural fluency. Schoenfeld also argues that to be effective problem-solvers, students need to be "familiar with a broad range of problem-solving strategies known as heuristics" (Schoenfeld 1985, p. 12). These are the rules of thumb for effective problem solving and broad strategies that help a student make progress on unfamiliar or difficult problems.

Adaptive reasoning in mathematics is about proving results, but it is also much wider. It involves the pursuit of what is true, usually leading to the formulation of a conjecture, testing it, and then trying to produce an argument that the conjecture is true. Extensive empirical or quasi-empirical testing of the conjecture may be required to verify that it is true, and would probably also involve a search for counter-examples (Lakatos 1976). This is where we see strategic competence working together with adaptive reasoning. Formulating an argument is similar to solving a problem and is required to convince ourselves and others of the truth of a conjecture (Open University 1997). Adaptive mathematical reasoning also can involve a search for *why* the conjecture or proposition is true. De Villiers (1990) argues that conviction that a statement is true very often provides the motivation for finding a proof. The purpose of such a proof is then to explain why the conjecture is true.

Ball (2003) hypothesizes that mathematical practices need to be deliberately cultivated and developed if they are to be acquired by students. Stein et al. (1996) look at what types of instructional environment might produce the desired mathematical thinking outcomes in learners. They suggest that students need to be exposed to "meaningful and worthwhile mathematical tasks" and that they are allowed sufficient time and given encouragement to explore their own mathematical ideas (Stein et al. 1996, p. 456). I believe that in this kind of classroom environment learners would develop the productive disposition strand of mathematical proficiency where they would "see themselves as effective learners and doers of mathematics" (Kilpatrick et al. 2001, p. 142). They would see mathematics as "useful and worthwhile" and would believe in their own ability to make sense of the subject (Kilpatrick et al. 2001, p. 131).

My Classroom and the Tasks

My pseudonym in this study is Ms. King. I conducted this case study in my grade 10 class, which was the second ability group out of a total of seven classes, so all 27 learners had achieved above average grades for mathematics. My school is an extremely well-resourced private boys school, with a teacher–learner ratio of 1:13 (see Chap. 2 for more details). Twenty of the boys in my Grade 10 class spoke English as their first language; the others spoke Afrikaans, Chinese or Japanese. There were no black learners in this particular class, although there are a few in the school.

I worked together with a colleague in this project to develop a set of tasks for our Grade 10 learners. I worked on these tasks with the class over approximately four lessons. The lessons were videotaped with a focus on what learners were saying as well as teacher–learner interactions. Unfortunately, a fifth lesson (task 5 below) was not captured on video because of timetable changes.

Although these tasks have been analysed in Chap. 2 in terms of cognitive demands, I present an analysis here to show how they could promote all five of Kilpatrick's strands. In planning the tasks, my underlying premise was borrowed from Stein et al., which stated that students need to "be provided with opportunities, encouragement, and assistance to engage in thinking, reasoning, and sense-making in the mathematics classroom" (1996, p. 457). I also needed to be aware of the cognitive demands of the tasks in order to match the tasks with my goals. In addition, I needed to scaffold the tasks appropriately, both in how I set them up, and in how I helped the learners to achieve the tasks without narrowing the task demands.

Task 1

Consider the following conjecture: $x^2 + 1$ can never be 0.

(a) Use a logical argument to convince someone else why the conjecture is either true or false for any real value of x.

(b) What is the smallest value of $x^2 + 1$? Explain how you know.

I expected that the learners would begin part (a) of this task by substituting positive numerical values for x. This would develop their conceptual understanding and procedural fluency when working with the concept of x^2 and then with the operation of $x^2 + 1$. They would need to execute this procedure accurately and appropriately for a representative sample of real numbers including fractions, negative numbers, and possibly irrationals. If they could not get to this stage where they applied concepts and performed procedures effectively, the rest of the exercise would be in jeopardy. This is one of the reasons why I chose groups for this work, as I believed that the learners could support each other around issues of procedural fluency and conceptual understanding. I expected them to be able to cope with the task at this level, even with negative number substitution, but perhaps with a little discussion and argument within their groups.

Through empirical testing, I expected the learners to convince themselves that the conjecture is true for any real value of x. They would need to use strategic competence and adaptive reasoning to produce a logical argument to convince someone else why the conjecture is true for all real numbers. I was hoping that some learners might produce a convincing algebraic argument. In part (b), the strategic competence and adaptive reasoning strands would be emphasised again. Learners would need to draw on their empirical testing and logical argument in part (a) to conclude that the smallest value is 1.

In terms of productive disposition, I hoped that the learners would work diligently at the task because it is interesting mathematically. Since this is a higher-level class, they do show interest in mathematical concepts, and I thought that the idea that you can have an expression that will always be positive would arouse their curiosity.

Task 2

(a) In the following list, the numbers on the right are related to those on the left:

x		y
1	→	1
2	→	4
3	→	9
4	→	16

(b) Can you find the rule that relates these numbers? Describe this rule in words.

(c) Can you write this rule mathematically?

(d) Ask your teacher to show you some other ways of writing this rule in mathematical notation. Then, describe in words what each of the different notations means.

This task was intended to be an introduction to functions and function notation. I chose this approach because it introduces the concept with the strategic competence strand and the ability to represent and solve a mathematical problem. I prefer this to the more frequently used approach wherein the concept comes first, then the procedure, and then the problem solving questions. In this way, I hoped to give the learners a glimpse of why we needed the concepts and procedures that were about to follow. In part (d), I wanted to introduce function notation as an alternative way to write $y=x^2$, and asked the learners to compare the different notations to support their understanding that there are different ways of writing the same concept in mathematics. To solve this task, learners would need to use strategic competence, conceptual understanding, and productive disposition, the latter because it is an unusual task for them and they would need to stay focused and see its importance.

Task 3

In order to talk about the above rule, we need to give it a name. We sometimes call it f, and we write it mathematically as $f(x) = x^2$. This means that when we apply the rule f to the number x, we get x^2. When we write $f(2)$, we mean "apply the rule f to the number 2". So $f(2) = 2^2 = 4$, Similarly, $f(3) = 3^2 = 9$.

Work out the following:

(i) $f(4) =$ (ii) $f(5) =$ (iii) $f(\sqrt{5}) =$ (iv) $f(-2) =$

(v) $f(-1) =$ (vi) $f(a) =$ (vii) $f(a + h) =$ (viii) $f(x + 1) =$

What do you think $f(a + h)$ means?

Definition of a function

We are now ready for a working definition of a function:

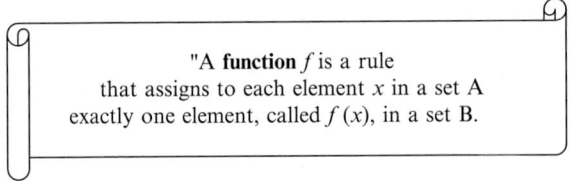

"A **function** f is a rule
that assigns to each element x in a set A
exactly one element, called $f(x)$, in a set B.

Explain, in your own words, what you understand a function to be. Draw a picture if this will help you to explain.

N.B.
We often represent functions with the letters: f, g, h. The letter in brackets after the f, g, h refers to the variable. We could write $f(x) = 2x + 1$, but would not write $g(x) = z + 3$, as the variables on the left and right hand sides do not correspond.

Task 3 continues with the learners reading and working out how to use function notation correctly and efficiently. They are then expected to practise using the notation and obtaining the correct answers, even though their level of understanding of what their answers mean is probably superficial. So at this stage, they are developing their skill to carry out the procedures accurately, working towards procedural fluency and conceptual understanding. The question "What do you think $f(a+h)$ means?" was inserted not to get a precise answer, but rather to encourage the learners to think about what their answers might mean and to have them think about what a function is to prepare them for the formal definition of a function that followed.

Task 4

If $g(x)=2x^2+3x-1$, evaluate the following:
(a) $g(1)$
(b) $g(-1)$
(c) $g(2)$
(d) $g(-2)$
(e) $g(a)$
(f) $g(-a)$
(g) $g(a+h)$
(h) $\dfrac{g(x+h)-g(x)}{h}$

I expected that the learners would start grasping the function concept and I wanted them to work with another example to consolidate their developing understanding. Question 4h was inserted to expose them to more challenging procedural work and to encourage them to work accurately.

Task 5

Consider the following statement: "$f(n)=n^2-n+11$ is a prime number for all natural numbers n."
(a) List the first 5 natural numbers.
(b) Determine $f(1), f(2), f(3), f(4)$, etc.
(c) Is the above statement true? Does $f(n)$ always generate a prime number?
(d) Try to justify/prove your answer in (c) above.

I chose this task because it demonstrates why empirical "proof" is not proof at all. It also shows that only one counter-example is needed to show that a statement is false. Part (a) provides the background for the task in reminding learners what natural numbers are. Part (b) reinforces procedures and shows a need for the function notation that they have just mastered. Perhaps this need is slightly contrived because the statement could have been written without the "$f(n)$," but I think that at grade 10 level, the $f(n)$ allows for a neat way of writing up their results. Part (c) promotes strategic competence as it requires learners to solve a mathematical problem, but it is in part (d) that the real mathematical reasoning is expected to occur. Learners are expected to reason adaptively as they need to explain mathematically why the formula fails. I hoped that this task would help them to see mathematics as sensible, useful worthwhile, and sometimes surprising, and so develop their productive disposition.

Initial Analysis

Classroom Interaction

My initial analysis of the videotapes was done by means of a coding system where I observed the video-recording of the lessons and marked off, at 2-min intervals, which of the first four strands (procedural fluency, conceptual understanding, strategic competence, and adaptive reasoning) was most evident. In order to help me to recognize the different strands, I expanded the definitions and listed, in detail, further descriptions of exactly what Kilpatrick et al. (2001) meant by the five strands. For example, in the conceptual understanding strand, the ability to explain the method to each other and correct it if necessary helped me to see Kilpatrick's et al.'s definition of conceptual understanding in practice. I excluded the productive disposition strand from my coding because it involves an attitude rather than an action, and this was difficult to infer from the videotapes.

In some of the 2-min intervals, more than one strand was evident. In such cases, I chose the strand that dominated. Another small problem that arose with the coding was that the strands might imply only correct contributions to mathematical proficiency. For instance, how would one classify learners making conceptual errors? Learners may be grappling with misconceptions and advancing their conceptual understanding even though they are making an error. I decided to include both correct and incorrect ideas in each strand if there was evidence that the learners were in fact grappling with what the strand requires.

I also experienced some difficulty with certain classifications. For example, in one case, I was unsure initially whether the difficulty the learners were having should be classified as conceptual or procedural. To help to resolve this problem, I conferred with the other members of the project team. We watched the relevant video-clip and reached a consensus that the learners were grappling with conceptual understanding. They knew how to execute both of the procedures correctly, namely $-(1)^2 = -1$ and $(-1)^2 = 1$, but they were confused about which one represented x^2 when $x = -1$. This consultation process gives increased reliability and validity to my study.

In order to establish the extent to which the four strands were present in the four lessons, I listed, on a spreadsheet, the two-minute time intervals for all the lessons. In the next column, I marked off the strand that dominated each time interval. This allowed me to determine the extent to which the strands had occurred in each lesson as well as over all the lessons.

The final breakdown of the four strands over the four lessons, given as a percentage of the total time is as follows:

Table 6.1 Strands in classroom activities

Conceptual understanding	41%
Procedural fluency	23%
Strategic competence	18%
Adaptive reasoning	18%
Total	100%

These results indicate that with carefully selected tasks and appropriate teacher scaffolding and interaction, it *is* possible to teach with all the strands working together. It is important to note that since these percentages are a quantification of qualitative data, the actual numbers are not that important, but rather the general trend. It is interesting to observe that the conceptual understanding category scored the highest. I was pleased that it was not procedural fluency because I believe that we need to move away from procedures dominating our mathematics teaching and learning. Conceptual understanding occurred approximately twice as frequently as the other strands, which were of a similar magnitude to each other. I was a little disappointed that strategic competence and adaptive reasoning did not feature more in my lessons, but they were present, which is important. As mentioned earlier, task 5 was not included in this analysis. If it had been, the strategic competence and adaptive reasoning strands may have been higher, because the final work that the learners handed in for task 5 included evidence of these two strands (discussed below).

The analysis of the video gave an indication of how much time was spent in class on the various strands. The next stage was to analyze the work handed in by the learners for evidence of the strands.

Learners' Work

For this analysis, I chose only tasks 1 and 5 as these were the two tasks that were intended to include all 5 strands. I took the work submitted by each learner and indicated which of the first four strands (excluding productive disposition) was evident in the written work. I argue that when learners used all four strands well, productive disposition would also be present. A summary of the results of the whole class for both tasks 1 and 5 follows.

In task 1, 14 (58%) of learners used the first four strands (conceptual understanding, procedural fluency, strategic competence, and adaptive reasoning) and completed the task fairly successfully. I believe that these learners also exhibited a productive disposition. A further 6 learners (25%) used all 4 strands to some extent, but were not completely successful in writing up the task. These learners showed some productive disposition characteristics because they worked steadily at the task but need more practice to develop their skill in writing down their answers. The remaining 17% did not use all 4 strands and did not complete the task successfully. Their work tended to show a lack of perseverance, so I have categorised them as not having productive disposition.

In task 5, 16 (73%) of learners used all 4 strands and completed the task fairly successfully. A further 5 (23%) used the 4 strands to some extent, but were not completely successful in writing up the task. The remaining one learner (4%) did not use all 4 strands and did not complete the task successfully. The results are summarized in Table 6.2 below.

It is interesting to observe that where learners were less successful in writing up the tasks, it was mostly because the strategic competence and/or adaptive reasoning strands needed more attention; generally their concepts and procedures were correct. However, the learners whose concepts and procedures were incorrect failed completely when it came to the strategic competence and adaptive reasoning.

Table 6.2 Evidence of strands in learners' work

	Task 1		Task 5	
	Number	Percent	Number	Percent
Comprehensive use of all 5 strands	14	58	16	73
Less comprehensive use of all 5 strands	6	25	5	23
Did not use all 5 strands	4	17	1	4
Total	24	100	22	100

It is also noteworthy that there was a general improvement from task 1 to task 5. This could have been because task 5 is easier, but on face value, it is a more difficult task, and it took the learners longer to complete the task. I believe that the learners improved because the whole class report-back and discussion on task 1 made them more aware of what was required in terms of mathematical reasoning and enabled them to understand the strategic competence and adaptive reasoning practices better.

Having given a quantitative overview of the strands in the lesson and the learners work, in the next section, I provide more detailed analyses of particular examples. This serves to illustrate how I made my classification to get to the above tables and also to illuminate the strands in practice more clearly for the reader.

The Five Strands in the Lesson

Procedural Fluency

The following extract comes from a teacher-led whole class discussion. It happened in the context of a question as to how to represent a negative number algebraically. Some learners had written $-x$, and I had spoken with them about having to consider that the value of x could be either positive or negative. To consolidate this concept, I decided to remind them of previous work on representing even and odd numbers algebraically. The learners would have covered this work before in both grade 8 and 9, and this was merely revision.

Ms King:	If I want to represent an even number, how would I do it in algebra?
Learners:	Two x
Ms King:	Yes, Evan
Evan:	Two x
Ms King:	Two x, Okay. And an odd number Evan?
Evan:	Two x minus one
Ms King:	Or
Evan:	Or add one, add one
Ms King:	Okay. Now, let's, we've thrown this x in, what does x mean?
Learners:	Any number, any real number

Ms King:	x can be what?
Learners:	Any number, any real number
Ms King:	Uh, can it
Learner:	No that was a mistake
Ms King:	For, for, for...Let's talk about even numbers. For even numbers, what can x be? can x be minus a half?
Learners:	No, Yes
Ms King:	Can x be point three?
Learners:	No
Ms King:	What can x be?
Learners:	Natural numbers
Ms King:	x can be a natural number; x is, and what are natural numbers Roelf?
Roelf:	One, two, three
Ms King:	x can be one, two, three, um...and here, same thing here. x is a natural number. So, just be careful. um, if you introduce an x, please define your x, say what your x means. In our old work, task 1, you were given x and knew what it meant. If you introduce an x, please define it; otherwise, I don't know what you talking about.

The above represents an example of procedural fluency because I was reminding the class how to represent even and odd numbers algebraically. Evan showed procedural fluency and/or conceptual understanding in reproducing the expressions. He could have remembered them from previous years or could have worked them out, based on his understanding of odd and even numbers. However, when the learners said that x could be any real number, I could see that they did not fully understand why the expressions represented even and odd numbers. Even when they finally said x must be a natural number, they might have been guessing, based on how they interpreted my question about x being $-1/2$. So there is some evidence that the learners did not understand the expressions but could repeat them.

Conceptual Understanding

The extract below shows a group of four learners working on task 1 wherein they had to agree or disagree with the conjecture that $x^2 + 1$ can never be zero. This group was struggling with how to represent x^2 when $x = -1$ and were arguing whether the minus sign should be inside or outside the brackets, i.e., whether $x^2 = (-1)^2$ or $-(1)^2$. Learners who were arguing for the latter were then claiming that $x^2 + 1$ can be zero, because x^2 could be -1. I came over to see what they were doing and to try to scaffold their thinking.

Ms King:	So when is it (the statement, $x^2 + 1$) false?
Roland:	When it's (he means x) minus one
Jimmy:	No, when it's minus bracket one bracket
Roland:	No
Jimmy:	If you put that in a calculator

Roland:	If x is minus one
Ms King:	If x is minus one
Roland:	But wouldn't that put it in a bracket automatically?
Jimmy:	No, not on a calculator
Roland:	Because if you take, like, x is equal to two squared, if you put that in a bracket as x. So you'll put the minus one in a bracket as x. So it won't work
Randall:	You can say minus one times minus one
Roland:	So if you say x is equal to minus one, it will be minus one there…
Jimmy:	No, x equals negative bracket one
Roland:	You have to put the brackets in because you have to put the whole term of x in
Jimmy:	x is equal to negative bracket one, like that
Randall:	You can't say that because then…
Roland:	No, you can't say that
Randall:	Then the whole thing would be negative x squared
Roland:	Jimmy, you can't say that. You have to put the whole thing inside brackets because the whole thing represents x. So it's (*writes* $(-1)^2$)
Randall:	Otherwise, the whole thing would be $-x^2$ (*writes*)
Ms King:	So when is it (*the statement, $x^2 + 1$*) false?
Roland:	When it's (*he means x*) minus one
Jimmy:	So if you put the whole thing in brackets and square it, it's going to be positive one

Roland and Randall were disagreeing with Jimmy as to how to represent x^2 when $x=-1$. Jimmy was working on his calculator, which was using algebraic logic and could not see the limitations of his representations. As discussed above, the learners' procedural fluency is not at issue here because they were all able to multiply both $(-1)^2$ and $-(1)^2$ quickly and efficiently. Their discussion was about which representation to use in this context, which is about conceptual understanding and perhaps some strategic competence. This task generated a lot of similar conceptual discussion in all the groups.

Strategic Competence

The following extract comes from the same group working on task 5. Is $f(n)=n^2-n+11$ prime for all natural numbers n?

Randall: Minus four squared is sixteen, minus minus four is twenty, plus eleven is thirty-one and thirty-one is a prime number.Jimmy: So we've got that one right. Okay, let's try minus one. If I do minus one, the answer I get is eleven. If I do minus three, the answer I get is twenty-three.
Randall: But negative numbers aren't natural numbers.
Jimmy: Yes, but we're trying to work out if it still works or not for non-natural numbers.
Roland: It does work for negative numbers.
Jimmy: Yes, but that's only one. Try minus one.
Roland: We've tried many. We've tried minus one, minus four. We've tried…
Rex: But why? We have to justify why.

I categorised this extract as strategic competence because the learners were finding ways to understand and solve the problem. In their attempts to find an answer, they were going beyond the limits of the problem as stated. This showed a flexible approach to problem solving. However, they had to remind themselves to come back to the problem constraints, and as we see in the last line, Rex was taking their problem solving to the next stage of adaptive reasoning.

Adaptive Reasoning

After working on task 1 (x^2+1 can never equal 0) for about 10 min, a group called me over:

Michael: Ma'am, it's simple ma'am.Ms King: Is it simple?
Michael: x^2 can never equal a negative number.
Ms King: Why?
Michael: Because if you square a number, if there are two negatives, minus minus, positive.
   If it's positive, it will be a positive.
Ms King: Yes, so…
Michael: It's always going to be a positive number. You're going to add something. It can never be zero.

I classified this incident in the adaptive reasoning category because Michael justified his conclusion using deductive reasoning. He worked from the definition of a square number and presented an explanatory argument as to why the expression must be greater than zero. His argument may not be complete in that he did not necessarily consider $x=0$, but he certainly had considered negative numbers. He was clearly convinced that he was correct and had managed to convince his group mates.

The Five Strands in the Learners' Work

The following example shows Jimmy's work where the first four strands were evident in task 1.

(a) x^2+1 can never be zero.

This statement is true.
For $x^2+1=0$, x^2 must be equal to -1, but the square of any number can never be negative. Therefore, x^2+1 can never be zero.

(b) The smallest value will be 1.

Since x^2 can never be a negative number, it has to be 0, or a positive number. 0 is always smaller than positive numbers. Therefore, $0+1=1$ is the smallest value for x^2+1, since $0^2=0$.

We can see evidence of Jimmy's conceptual understanding of the idea of squaring a negative number. Although his procedural fluency is not fully demonstrated, it is possible to infer that he would have needed it in order to calculate values for the expression. This is somewhat evident in part (b). From my analysis of the video (see above), I know that Jimmy initially made a conceptual error in relation to the difference between $(-1)^2$ and $-(1)^2$, but he had sorted this out by the time he wrote up his solution. He devised an appropriate solution method appropriate for the problem, showing strategic competence and used adaptive reasoning to provide a convincing argument.

This next task shows Clive's final submission for task 1, which scored zero for work in the conceptual understanding, procedural fluency, and adaptive reasoning strands. There is a little strategic competence evident because Clive tried to solve the problem but with inadequate concepts. He was unable to reason adaptively because the information that he was using was based on a flawed concept.

(a) It is true or false.
     True if x=any no.
     but false if -1 isn't in brackets.
     e.g., $x=2$, $2^2+1\neq0$
     $-1^2+1=0$
     $(-1\times1\times1)+1=0$

In response to task 5, this next example shows all the strands:

(a) 1; 2; 3; 4; 5
(b) $f(1)=1^2-1+11=11$
     $f(2)=2^2-2+11=13$
     $f(3)=3^2-3+11=17$
     $f(4)=4^2-4+11=23$
     $f(5)=5^2-5+11=31$
     $f(11)=11^2-11+11=121$
     121 is not a prime number
(c) No/false. Multiples of 11 don't work.
(d) Natural no. are from 1… onwards. 11 is a prime number. $\therefore$ as we see above $11+11$ cancel out and $=n^2$ and n^2's factors are 1, n and n^2. $\therefore$ 3 factors=non-prime

This work showed conceptual understanding and procedural fluency because the learner knew what the natural numbers are and how to substitute them correctly using function notation. He also knew how to identify the answers as prime or not. He showed good strategic competence because he established that multiples of 11 do not work. Adaptive reasoning and strategic competence were evident in part (d) when he justified his answer by showing that, when you substitute 11, the result is the square of a natural number, which can never be prime.

Conclusion

So, is it possible to teach and learn in a classroom where all five strands are emphasised and synchronised? The tasks were selected carefully for their potential to promote all of the strands, and the analysis of the lessons allowed me to conclude that they were effective in doing so for most of these learners for the following reasons:

1. The videotape coding of the classroom activities indicated that in terms of time, all strands were in fact present to a significant extent, although I would have preferred more strategic competence and adaptive reasoning to be evident in the classroom.
2. My analysis of the learners' work indicated that approximately 75% of learners used all five strands to some extent in task 1, and approximately 95% used all the strands in task 5.
3. I attributed the improvement in quality of work submitted to the fact that the learners had acquired a better understanding of what was required. They achieved this by participating in whole-class, teacher-led discussions, which enhanced their ability to work with all the strands.

The results of this research may not be generalizable to all grade 10 classes because this particular class was above average in terms of mathematical achievement and was from a well-resourced school. However, from the positive results of this small research, I am encouraged to experiment further with different classes as I now believe more strongly that it is possible to teach and learn in an environment where all five strands are present and reinforce each other.

If other teachers were to attempt similar research in their classes, the critical factors that need to be considered are that the tasks need to be cognitively demanding in terms of the kind and level of thinking required (Stein et al. 2000). Learners should also be exposed to such tasks on a regular basis in order to develop strategic competence, adaptive reasoning, and productive disposition more fully. Based on my experiences, I would recommend that teachers take the plunge and begin to teach in this way. It is not easy, but careful thought and planning can help, and it is very rewarding to feel that you are teaching in ways that are recommended, not only by the new curriculum in South Africa, but in other countries as well.

Chapter 7
Teaching the Practices of Justification and Explanation

Mathematics is generally perceived to be a difficult subject and as Ball (2003) states there is a "cultural belief that only some people have what it takes to learn mathematics" (p. 34). My motivation for researching mathematical thinking and reasoning stems from wanting to change this perception among my learners. From my reading, I am encouraged in thinking that this perception can be changed. Lampert (2001) and Heaton (2000) argue that changes in pedagogy towards teaching that promotes mathematical reasoning can afford all learners the opportunity to increase their thinking and reasoning abilities, and change the perception that only the selected few can do mathematics.

In my teaching, I have been frustrated by the silence that I am often greeted with when I ask learners to explain their reasoning. Learners, and even parents, are of the opinion that it is the teacher's job to teach and all they have to do is to copy notes and complete exercises. When trying to encourage justification and explanation, learners often resort to the ever present "Sir, just tell us". Ball et al. argue, "discouraged teachers may conclude that these students cannot do more complex work and may return to simpler tasks" (2003, p. 35). Therefore, they argue that "a focus on the practices for learning supply more learning resources needed by teachers and learners to engage in more ambitious curricula and working toward more complex goals" (2003, p. 35).

With the dawning of a new curriculum in mathematics for Grades 10–12 (Department of Education 2003), teachers in South Africa are encouraged to adopt new approaches to teaching and learning. The National Curriculum Statement envisages that learners need to work together in solving challenging problems. This will provide them with an important basis for acquiring a wide range of personal skills for the future by promoting creative thinking, improving communication and co-operation skills, and strengthening learners' ability to acquire new knowledge.

This study sheds some light on the pedagogic demands, challenges, and possibilities of the new curriculum. In working in an under-resourced school, I show how learners can be given opportunities to interact with challenging tasks, develop and demonstrate their reasoning skills and creative thinking, form generalizations, make conjectures, and justify their claims. Through an analysis of my teaching, I show how teachers can facilitate this process. I will also show that while it is

K. Brodie, *Teaching Mathematical Reasoning in Secondary School Classrooms*,
DOI 10.1007/978-0-387-09742-8_7, © Springer Science+Business Media, LLC 2010

possible to achieve some success in this direction, this kind of teaching throws up immense challenges for teachers.

Construction and Practices

This study is informed by a constructivist theory of learning. To explain constructivism, Hatano (1996) argues, "Humans often explore tasks beyond the demands or requirements of problem solving and that we "try to find 'meaning' or a plausible interpretation of their observations of facts and effective procedures" (p. 61). So constructivism suggests that we all interpret and enrich the knowledge and experiences that come from interacting with our environments, both social and physical. Constructivism also argues that as learners interpret new knowledge, they integrate it with their current knowledge and in so doing, restructure and reorganize their current knowledge (Hatano 1996). This means that the learner her/himself must make sense of new ideas; no one can do it for her/him. This is why constructivism forms a basis for learner-centred pedagogies (Brodie et al. 2002). At the same time, this does not mean that there is no role for the teacher. Hatano (1996) argues that social influences, including the teacher's input, are central in supporting learners' constructions.

The Ball (2003) defines mathematical practices as "the things that mathematicians and mathematical users do" (p. 24). Examples of these practices include using symbolic notation, justifying claims, and making generalizations. I am particularly interested in the practices of justification and explanation. These two practices form an integral part of a constructivist perspective because as learners explain and justify, they are able to construct new meanings and build on their own and each other's thinking. Lampert (2001, p. 64) argues that "implicit in these [the learners] responses is the message that students can and do give explanations, and can and do build on one another's thinking". In order for learners to build on one another's thinking, new norms (McClain and Cobb 2001) need to be taught to create a classroom culture for explanation and justification. This classroom culture should include learners' tolerance and respect for others' contributions or ideas, being polite within disagreement and building on each other's contributions.

The Practices of Justification and Explanation

Davis (1988) makes a provocative statement: "Mathematics, as it appears in most classrooms, is a weird thing. There is virtually no mathematics in it" (p. 1). He observes that the mathematics taught is only a collection of rituals to manipulate symbols. He argues further that these rituals are meaningless and the students see symbols that do not make sense. Rituals like "Take all x's to one side and the numbers to the other" are emphasized with no reasons given.

Skemp (1976) distinguishes between two types of mathematical understanding: instrumental understanding and relational understanding. He describes relational understanding as "knowing both what to do and why" (Skemp 1976, p. 23), whereas instrumental understanding is "rules without reason" (p. 16), which requires "a multiplicity of rules rather than fewer principles of more general application" (p. 19). Instrumental understanding is characterized as being "easier to understand, the rewards are more immediate and the right answer comes quickly" (p. 22). Skemp lists the following as advantages of relational mathematics: it is more adaptable to new tasks; it is easier to remember because it puts ideas into relation with each other and into a broader framework; the quest for relational knowledge is often a goal in and of itself; and relational knowledge is organic, i.e. it can create possibilities for growth of more knowledge (p. 22).

The new curriculum in South Africa encourages the development of relational understandings. The National Curriculum Statement (Department of Education 2003) includes as part of its definition of mathematics that, "Knowledge in the mathematical sciences is constructed through the establishment of descriptive, numeric and symbolic relationships. Mathematics is based on observing patterns; with rigorous logical thinking this leads to theories of abstract relations" (p. 9). The argument is that learning to explain and justify mathematical reasoning supports the construction of relational knowledge. Davis (1988) concurs by stating that "written symbols on paper, together with the ever present meaningless rules will never add up to mathematics" (p. 2).

Kilpatrick et al. (2001) strengthen this argument with their notion of mathematical proficiency that is constituted by "five, interwoven and interdependent strands" (p. 116). These have been discussed in detail in the previous chapters. The important point for this chapter is that relational understanding, or conceptual understanding in Kilpatrick et al's terms, depends on the development of justification and explanation. My research is located in the strand described as adaptive reasoning, which is "the capacity to think logically about the relationships amongst concepts and situations" (Kilpatrick et al. 2001, p. 129). They argue that "such reasoning is correct and valid, stems from careful consideration of alternatives, and includes knowledge of how to justify the conclusions" (Kilpatrick et al. 2001, p. 129). Kilpatrick et al. (2001) argue that adaptive reasoning is much broader than formal explanation and justification, and includes intuitive and inductive reasoning based on pattern, analogy, and metaphor. These they describe as the "tools to think with" (p. 129) to aid learning.

My investigation of the mathematical practices of justification and explanation is also guided by the research conducted by Ball (1993). He argues that through a study of mathematical practices, researchers can describe what learners need to know and do to become mathematically proficient, for example, to symbolize, represent, justify, and explain. In order to help learners to achieve these mathematical practices, it is important to understand the contributions made by tasks and teacher intervention strategies. My interest in this project was to adapt my teaching approach so that learners would be presented the opportunity to justify and explain their mathematical claims. I also wanted to research the ability of my learners to mathematically explain and justify their thinking and reasoning. Ball (2003) argued that teaching with an emphasis on practices can support "low-achieving learners who may have had less opportunity to develop these practices [to] achieve significant gains"

The Importance of Tasks

Stein et al. (2000) state: "some tasks have the potential to engage students in complex forms of thinking and reasoning while others focus on memorization or the use of rules or procedures" (p. 14). They argue that if the goal for learning is to increase the ability to think, reason, and solve problems, then students need to start with cognitively challenging tasks that have the potential to engage students in complex forms of thinking. Boaler (1997), in her research in two schools, Amber Hill and Phoenix Park, illustrated the effects different tasks had on students' ability to think, reason, and solve problems. While the students at Amber Hill engaged in "cue based learning", those at Phoenix Park were exposed to "open ended tasks", which supported the construction of flexible knowledge. Boaler (1997) confirms Stein et al's. (2000) argument , that focusing on tasks that support procedures only, without connections to underlying concepts and meanings, can lead to a limited understanding of what mathematics is and how one does it. Boaler (1997) argues:

> Although the Amber Hill students spent more time on tasks in lessons and completed a lot of text book work, it was the students of Phoenix Park who were able to use mathematics in a range of settings" (Boaler 1997, p. 116)

Stein et al. (2000) classify tasks based on the kind of thinking they demand of students, i.e. the cognitive demand of the task. Tasks that require a mathematical explanation or justification are described as procedures with connections tasks. Tasks, for which no pathway is suggested, i.e. "there is no overarching procedure or rule that can be applied for solving the entire problem and the sequence of necessary steps is unspecified" (p. 15) are classified as doing mathematics tasks. These two kinds of tasks are high-level tasks (see Chap. 3 for more detail). These tasks tend to be the ones that require explanation or justification from learners.

The Teacher's Contribution

As was argued in Chap. 1, high-level tasks are not sufficient to engage learners in the practices of justification and explanation. So an important part of this study was to understand whether and how my teaching could promote explanation and justification. The new curriculum in South Africa redefines the role of the teacher to that of a facilitator. However, what this means is often not well specified (Brodie 2003). Many people believe that to be a facilitator, a teacher must not tell anything to the learners, but should rather question them towards an understanding.

Chazan and Ball (1999) argue that telling teachers "not to tell" does not go far enough because it does not tell teachers what they should be doing. Brodie (2007a) argues that questioning does not always challenge learners; in fact, it can narrow and funnel their responses (Bauersfeld 1988). For this study, I found the skills and knowledge outlined by Schifter (2001) to be very helpful.

1. *Attending to the mathematics in what one's students are saying and doing*
New methods of teaching ask teachers to listen more carefully to learners, what Davis (1997) calls interpretive listening – i.e. listening *to* the learners, rather than *for* what the teacher wants to hear. What is important here is to listen to the particular mathematical ideas, correct or incorrect, that learners express in order to understand their thinking.

2. *Assessing the mathematical validity of students ideas*
This skill builds on the first. As teachers listen interpretively to the mathematics in learners' contributions, they need to assess the possibilities for using the contribution to take the learner's and the class's thinking forward. This is different to listening for checking whether the answer is correct or not. The teacher should also encourage learners to assess the validity of their own and each other's solutions. It is critical for the teacher to listen to mathematical arguments and assess the validity of those arguments. In the process, teachers will develop their own powers of mathematical thought.

3. *Listening for the sense in students mathematical thinking even when something is amiss*
Teachers need to search for the sense in what learners are saying or doing. They can look for mathematical strengths or for learning opportunities. They can work through learners' logic and try to appreciate their sense making, even if there seems to be no logic. This skill builds on understandings of constructivism and learner-centred teaching – that whatever the learner says, no matter how incorrect, must somehow make sense to her/him. In seeking the learner's mathematical logic, the teacher discovers the work that needs to be done.

4. *Identifying the conceptual issues the students are working on*
Mathematical errors that are uncovered during a lesson can be made public for the class to consider and discuss (see also Swan 2001). Teachers need to identify what mathematics is missing, to see the logic of the learners' thinking and then try to identify the general mathematics concepts with which learners are finding difficulty. Having done so, the teacher can proceed by rephrasing the problem or posing a question to help the learners to see the situation in a new way or in another context.

Up until this point, I have drawn on a range of literature to support my study. Drawing on the above literature, this study is anchored in a number of principles:

- Learners are at the centre of the teaching process;
- Teachers are central in supporting learning even though they can't learn for learners;

- All learners can construct their own 'new' knowledge through making mathematical meaning and formulating, testing, and justifying conjectures.
- All teachers can develop skills to support learners' justification, reasoning, and meaning making.

My Classroom

In this study, my pseudonym is Mr. Peters. My school is in western Johannesburg and is a dual-medium school, with English and Afrikaans being the languages of learning. The school is situated in a poor socio-economic community infested with gangs and drug trafficking. The school is poorly resourced, with old style desks, classrooms in disrepair, and no regular electricity in classrooms. There is paper for printing worksheets, but there are few textbooks for learners. The subjects of this study were learners in my grade 10 class, of 45 learners (boys and girls). The class is of mixed mathematical ability, although as discussed in Chap. 2, most of the learners have weak mathematical knowledge. Learners' ages range from 14 to 16. The learners came mostly from working class families, or their parents are unemployed. Many learners live with grandparents or other relatives. The language of learning in this class is English; however, the main languages of the learners differ considerably and include English, Afrikaans, isiZulu, seSotho, and xiTsonga.

Although I had not taught any of these learners previously, from my knowledge of other teaching in the school and from their responses to my initial requests for explanation and justification, I speculated that their contact with higher-level tasks was minimal or non-existent. My primary goal in this research was to engage learners in proving conjectures through the mathematical practices of justification and explanation. I was inspired by Lampert's goal that "everyone in the class was to study mathematical reasoning" and her role was to "teach them (learners) new ways to think about doing mathematics and what it means to be good at mathematics" (Lampert 2001, p. 65). In addition, an important goal for teachers is to use mathematical reasoning to teach content. As I worked with the learners on these tasks, I paid attention to developing in myself, Schifter's four skills as discussed previously.

I worked together with a colleague in this project, to develop a set of tasks for our Grade 10 learners. However, since my learners were a lot weaker than my colleagues, and since their experience of these kinds of tasks was minimal, I changed some of the tasks in order to scaffold the process more carefully. Although I anticipated the difficulties my learners would experience and planned the tasks around these difficulties, at no stage did I simplify the tasks or "dumb them down". I was very aware of the fact that in Stein et al.'s research "only about one third of the tasks that started out at a high-level remained that way as the students actually engaged with them" (Stein et al. 2000, p. 24) and did not want this to happen in my classroom. I worked on these tasks with the class over four lessons, each one approximately 45 min. long. The lessons were videotaped with a focus on what learners were saying as well as teacher–learner interactions, which enabled me to do a detailed analysis of parts of the lessons. As learners worked on the tasks in their groups, they were asked to write a final response. These written records were also useful to analyze, to try to understand the learners' developing practices of justification and explanation.

The tasks have been analyzed in detail in Chap. 2. In this chapter, my analysis focuses on Task 1(B), so I will repeat it here. Since this was the main focus, the other parts of task 1 were intended to scaffold this task (see Appendix).

Task 1(B):

Consider the following conjecture: "$x^2 + 1$ can never be zero"

Prove whether this statement is true or false if $x \in R$.

The task is open-ended, and no suggested pathway is offered to the learners. In choosing the task, I believed that it would engage learners in complex forms of thinking and reasoning. Using Stein et al's (2000) framework, I would classify it at the highest task level of "doing mathematics" as it offers a variety of solutions to the problem (see also Chap. 2). My ideal solution for this task was that learners would recognize that x^2 is a perfect square and therefore has a minimum value of zero for all real values of x and when added to one, this will result in a positive value. Hence the conjecture that $x^2 + 1$ can never be zero is true. Another solution would be to say that for $x^2 + 1$ to be zero, x^2 must equal -1. This implies that $x = \sqrt{-1}$, which is impossible as we are working with real values for x; therefore $x^2 + 1$ will never be zero. I did not expect my learners to work with imaginary numbers. A third solution would be a graphical representation of $y = x^2 + 1$, which would display that there is no x-value to satisfy the equation when $y = 0$. I also expected some learners to substitute values to see what would happen to the expression. As will be seen in the analysis below, many of my learners did not even get close to these solutions, although some did make some progress. I had to work very hard to develop appropriate justifications among most of the learners.

I gave the learners 40 min. for this task so that they would have enough time to discuss the conjecture with their partners. We then had an extended whole-class discussion, based on report-backs from their groups. An analysis of their written responses and the whole-class interaction follows.

The Learners' Written Responses

Twenty pairs of learners handed in their responses at the end of the session. I first classified the responses in relation to those who said the conjecture was true and those who said it was false. Only one pair said that the conjecture was false, their reason being that x can be zero (see example 1 below). For the learners who said the conjecture was true, I further classified their responses according to whether their justification was incorrect, partial, or complete. Table 7.1 gives an overview of these responses, and following Table 7.1 are examples of the different kinds of justifications.

Table 7.1 Justifications for the conjecture being true

Incorrect justification	$x^2+1 = x^2+1$ because it can't be simplified	13 pairs
Partial justification	Adding a sum, plus another value will get a higher value, not zero	2 pairs
	Substituted positive values and zero for x. No negative values considered	3 pairs
Complete justification	Substituted positive, zero, and negative values for x. Realized that x^2 must equal -1 for conjecture to be true	1 pair

1. False – incorrect justification

false, because   x is any real ruber and zero
1. is a real number x can just take the fome of
zero

2. True – incorrect justification

x^2+1 can never be Zero., True, because it can never be zero; $x^2+1 = x^2+1$ because
Cannot add unlike or subtract unlike terms. The prove is x^2+1 it will still give
you x^2+1 because x^2 is an know value and 1 is a Real number. They cannot be
added because $x^2+1 = x^2+1$ and it wont give you any other anwer

$$x^2+1 = x^2+1$$

3. True – partial justification

¥ True because when you find the value of x^2
plus by one you n number which is not zero
eg $2 \times 2 + 1 = 5$

   0×0 +: →

   $5 \times 5 + 1 = 26$

   10×10

* it would be zero because you add one ever
if you take one

4. True – complete justification

Its true because x can be any
number Squared and when the x^2
is added to the number it will give you
a Sum

$= 2^2 + 1$

$= 4 + 1 = 5$

$x^2 + 1$

$= 0^2 - 1 1$

$= 0 + 1$

$= 1$

$x^2 + 1$

$= -3^2 + 1$

$= +9 + 1$

$= +10$

$x^2 = -1$

Table 7.1 shows that the majority of learners gave an incorrect justification, using an algebraic argument that since the expression could not be simplified, it could not equal zero (example 2 above). While this justification made sense to me in terms of learners' prior knowledge, it concerned me because it showed that learners did not recognize the fact that if they substituted values for x, the expression would have a value. This means that learners' procedural knowledge was static, not flexible, because it was not connected to conceptual understanding (Kilpatrick et al. 2001).

Five pairs of learners gave partial justifications. Two pairs argued that the terms could be added if they were given values, and that adding two values would give an answer greater than zero. These responses showed some intuitive reasoning; however, they were poorly expressed and did not show that the learners actually understood that x^2 could not be less than zero. Three pairs substituted numbers in the expressions (example 3 above). This approach is promising if extended to negative numbers and a more theoretical explanation; however, these learners did not get to that point. Only one pair gave what I considered to be a complete justification (example 4 above). This pair had started by testing numbers.

With the above analysis, I returned to the classroom the next day, concerned that the majority of my learners did not understand the conjecture or know how to justify it. There were a number of questions that needed to be asked regarding their mathematical reasoning:

- Why did the majority of learners see $x^2 + 1$ as unlike terms and not as a variable expression which could have a value?
- Did they realize that $x^2 + 1$ is a unit value if x is known or did they see them as unlike and therefore separate terms?
- Did their limited conceptual and procedural understandings limit their possibilities for adaptive reasoning and justification?

I hoped to generate discussion among those who had given incorrect justifications and those who had given partial or complete justifications, so that they could engage with different ways of seeing the problem. I therefore selected a response in each of the three categories in Table 7.1 to discuss in class the next day. The first response came from Grace and Rethabile (example 2 above), as typical of those who gave incorrect justifications. They were representative of a large group of learners who had a shared view that I needed to know more about. The second pair of learners, Jerome and Precious (example 3 above), was selected because they used a systematic, empirical method of testing the conjecture. Unfortunately, they had not moved on to a complete justification. The third pair, Marlene and Tshepo, (example 4 above) was the only pair who had formulated a correct explanation to justify the conjecture that "$x^2 + 1$ can never be zero". I was impressed by their reasoning ability and wanted to afford them the opportunity to share it with the class.

Whole-Class Interaction

My reasons for having the class discussion about the learners' written responses were threefold. Firstly, I needed to understand learners' reasoning in their written responses. Were they understanding more than what they wrote? Had they more to offer than their response on the page? Was I reading and interpreting their written responses correctly? The second reason for having the discussion was to establish whether learners could reflect on their own thinking and build on it, especially to transform misconceptions and partial reasoning into successful understandings and reasoning. The third reason was for the class to understand what other learners were thinking and reasoning and to engage them in a discussion, hopefully also to develop their explanation and justification practices.

Incorrect Justification

I began the lesson by writing up Grace and Rethabile's solution and asking the two girls to comment on it. I did not give any indication as to whether they were correct or not. Rather, I invited them to say more about what they were thinking so that I could attend to the mathematics in what they were saying and doing (Schifter's first skill).

Mr Peters:	Grace and Rethabile. And most of you belong in this group. (*writes Grace and Rethabile's solution*) Grace, do you want to say something about that? What were you thinking? What were the reasons that you *(inaudible)*
Grace:	Sir, because the x squared plus one ne sir, you can never get the zero because it can't be because they unlike terms. You can only get, the answers only gonna be x squared plus ones that's the only thing that we saw because there's no other answer or anything else.
Mr Peters:	How do you relate this to the answer not being zero? Because you say there it's true, the answer won't be zero, because x squared plus one is equal to x squared plus one. You say they're unlike terms. Why can't the answer never be zero, using that explanation you are giving us?
Grace:	*(sighs and pinches Rethabile)*
Mr Peters:	Rethabile, do you wanna help her
Rethabile:	Yes, sir.
Mr Peters:	Come, let's talk about it.
Rethabile:	Sir, what we wrote here, I was going to say that the x^2 is an unknown value and the 1 is a real number, sir. So making it an unknown number and a real number and both unlike terms, they cannot be, you cannot get a zero, sir. You can only get x squared plus one.
Mr Peters:	It can only end up x squared plus one
Rethabile:	Yes, sir. There's nothing else that we can get, sir. But the zero, sir.

In the transcript, Grace repeated exactly what she had written, that x^2+1 equals x^2+1 since they are unlike terms. She made it quite clear that the only answer we will get if we add x^2 and 1 is x^2+1. She re-emphasised her misconception and did not challenge it. My next move, to ask her to relate her explanation to the answer not being zero was an attempt to apply Schifter's skills 2 and 3. I wanted to listen to their thinking and try to assess the validity of their mathematical argument and make mathematical sense of what they were saying. I also hoped to encourage them to do the same. However, Grace did not want to respond to my move and called on her partner, Rethabile, to help her, which I re-inforced.

Rethabile was more explicit in saying that x^2 is an unknown value and 1 is a real number. Therefore, the unlike terms cannot be added. This could help for the later discussion, but Rethabile could not take it further, insisting that they could not get any other value for the expression. As a teacher, I could understand their explanation. It was true that x^2 and 1 were unlike terms, as they had learned previously. However, they seemed unable to move beyond this idea and see that if x is given a value, $x^2 + 1$ will have a value as well, i.e. if you substitute a real value for x, you do get a number for x^2, which you can add to 1. My challenge was to decide how to work with their misconception in a way that would help them to move beyond it.

My next move was to call on other learners to confirm or challenge what Grace and Rethabile had said. Many other learners had used a similar justification to argue that the conjecture was true.

Mr Peters:	Come there are many of you who wrote x squared equal to one. Wrote unlike terms. Who wants to add to what she is saying? What were you thinking, … shes saying that, yes Ahmed
Ahmed:	Like you always say sir, in front of the x you see a one.
Mr Peters:	In front of the x you see a (*pause*)?
Learners:	one
Mr Peters:	A one, here.
Ahmat:	So um, you plussing one still, sir
Mr Peters:	But this
Ahmed:	You can get an answer.
Mr Peters:	So, what is the answer
Ahmed:	Um, if you have one x squared plus one, sir.
Mr Peters:	What's one x squared plus one
Ahmed:	If I add it, if I add it sir, one x squared would give me, plus one would give me two x squared.
Mr Peters:	Would give you two x squared. Right (*pause*), so x squared plus one, because of the one there, Ahmed says, that's gonna be equal to two x squared. [*writes on board* $1x^2 + 1 = 2x^2$] Has anyone got something to say about this?
Learner:	Yes, sir.
Mr Peters:	Victor
Victor:	x squared plus one stays the same because you can't add the unlike terms, so if x squared plus one s equals to two x sqaured, because if
Mr Peters:	So you agree with, sorry, you agreeing that x squared plus one is equal to two x squared?
Victor:	Yes, sir. If I add my one's sir
Mr Peters:	Which one's are you adding?
Victor:	In front of …
Mr Peters:	This one and that one?
Victor:	In front of x, sir. I chose 1 because x has unknown value so I prefer to take one, sir. And I make it one x squared plus one.
Mr Peters:	So you adding one plus one to give you the two there?
Victor:	Yes, sir. (Some others also say yes)

Ahmed brought out a misconception that many learners were struggling with. I felt it necessary to discuss this misconception, as there were other learners with the same view as Ahmed. I was interested in what these learners were saying and reasoning. Were they saying that x^2 is 1 because of the 1 before the x^2? Here I was using Schifter's skills 3 and 4, trying to make sense of the learners' incorrect thinking and deciding on the conceptual issues that they were working with or needed to work with. Schifter states that conceptual errors uncover a deep mathematical issue for the class to discuss. I agreed with Schifter that this learning opportunity to clear up a conceptual issue should not be ignored, so my next move was to open this issue up for class discussion. Victor made a somewhat contradictory statement, repeating both Grace's point that you can't add unlike terms and Ahmed's conviction that x^2+1 would give $2x^2$. Victor's idea was slightly different from Ahmed's, suggesting that the 1 in front of the x^2 could be added to the other 1.

My next move was to call on Lebogang whose body language, raising her hand, and shaking of her head while Ahmed and Vincent spoke, suggested that she had a different idea from them. Lebogang disagreed vehemently with her fellow learners and supplied an adequate justification for her disagreement.

Lebogang:	Um, I disagree because one x squared plus one, they are unlike terms because of the x squared. They are unlike terms, you can't add unlike terms.
Learner:	Ja
Mr Peters:	You can't add unlike terms.
Lebogang:	(and some others) Yes.

After this, I tried to determine the effect of Lebogang's contribution. Had Lebogang's contribution and justification help other learners to shift their thinking? Had Lebogang provided sufficient mathematical reasoning to convince Victor and Ahmed differently? I determined that about half the class agreed with Ahmed and the other half with Lebogang. So I reminded them why they could not simplify unlike terms. However, this left me reinforcing the incorrect justification for the original task. At this point, I felt that the best way to deal with this was to discuss the solution of a pair of learners who had substituted values for x.

Partial Justification

Jerome and Precious had started by substituting two for x, which resulted in $x^2+1=5$ and then moved on to $x=0$, which gave $x^2+1=1$. I questioned them on their choice and Jerome said that they used any number.

Mr Peters:	What, what, what were you doing there Jerome?
Jerome:	I said x was equal to two, sir
Mr Peters:	Here, you said x is equal to (pause)
Jerome:	Two
Mr Peters:	x is equal to two. Why did you choose two? Why didn't you choose any other number?
Jerome:	I just took any number.
Mr Peters:	You took any number. Okay.

I accepted this, trying to "listen to" what Jerome was saying, rather than listening "for" what I wanted to hear (B. Davis 1997). I proceeded by asking them: why zero?

Jerome:	I said x is equal to nought, sir.
Mr Peters:	You said x is equal to zero.
Precious:	Yes, sir.
Mr Peters:	Why did you choose zero? First you chose two, now you're choosing zero. Why?
Jerome:	I chose any number sir.
Mr Peters:	I want to believe you are choosing it for a reason and not just taking any number.
Jerome:	I was trying to get out, to see if it ends up on nought, sir.
Mr Peters:	You were trying to go closer to (pause)?
Jerome:	Nought, sir.

I put Jerome and Precious' work up for discussion because I believed that there was a learning opportunity for the class as a whole. There was mathematical sense in what they were doing, a systematic empirical testing on values for x; however, there were limitations to their method that I needed to clarify. Was their method of solution complete and did they think it was complete? Here I tried to assess the validity of their thinking and also get them to understand their own thinking. However, Jerome and Precious did not respond with an adequate explanation of their thinking. Jerome argued that they did not choose zero for any particular reason. At that point, I had to intervene, insert my own voice (Chazan and Ball 1999) and say that I thought they were choosing it for a reason. I had to bring out the reasoning in their thinking, because they did not articulate it themselves. I wanted to show them and the class that by systematically moving towards and beyond zero in testing the numbers, they could begin to see a pattern. I did this in order to move the class forward (Heaton 2000). I worked with Jerome's thinking, amplifying it for the class, and showing the reasoning that I assumed was behind it.

After this, I asked: What is the next logical value to substitute for x and why? Marlene immediately responded: negative one, and I decided to give her the opportunity to explain her and Tebogo's solution.

Correct Justification

Marlene went up to the board and interacted with the class as follows:

Marlene:	What's negative one times negative one?
Class:	Positive one (*she writes* $-1x-1 = +1 + 1 = 2$) Negative one, positive one
Marlene:	(*writes:* $-1x-1$) You'll never get to zero sir
Ahmed:	You will get it
Class:	You won't
Marlene:	Unless
Mr Peters:	OK, give Marlene a chance
Marlene:	Unless it was negative one times positive one
Class:	*shouts inaudibly*

(continued)

Marlene:	Then you'd have plus one, a negative one and plus one will give us nought
Class:	Yes
Marlene:	But sir we have x squared which means we must have two numbers which are identical
Class:	Yes
Learner:	It won't give zero, it will never give zero, because that's like terms

When they worked together the previous day, Marlene and Tshepo had derived a complete response with a correct mathematical explanation and justification. They, like Jerome and Precious, had started by substituting values for x; however, they substituted positive values, zero and negative values and saw that each time $x^2 + 1$ did not result in zero. But they did not stop there. They proceeded to find a mathematical explanation to justify their claim and concluded that for $x^2 + 1$ to be equal to zero, x^2 must be -1 as $-1 + 1$ equals 0. But x^2 cannot be equal to -1, as we cannot find the square root of -1.

In the interaction above, Marlene used a different explanation, possibly influenced by the previous discussion with Jerome. She showed, using -1 as an example, that multiplying two negatives will give a positive. When Ahmed disagreed with her, she rethought her explanation, suggesting that only if you had -1×1 could you get zero. She then argued again that -1×1 is not an appropriate representation of x^2. So even though Marlene had a correct, convincing argument, having to explain it and being challenged by another learner, forced her to rethink her idea and come to the same conviction, possibly strengthening it. However, at that point, I did not "celebrate" her solution (Lampert 2001, p. 160), because I wanted the class to engage in reasoning about it. My next move was to question Marlene's correct answer. Lampert (2001, p. 53) argues that teachers should promote learners to be generators and defenders of mathematical knowledge. Teachers need to send a clear message to learners that all answers, correct or incorrect need to be justified. The learners' mathematical reasoning regarding the issue is more important than the answer itself.

Mr Peters:	But what have we got here? We've got an x squared. So the x squared must be equal to, Marlene
Marlene:	Negative one
Mr Peters:	Negative one
Marlene:	But there is no square root of negative one.
Mr Peters:	What d'you mean there's no square root of negative one.
Marlene:	Sir, can you tell me what the square root of negative one is?
Mr Peters:	The square root of negative one is negative one.
Class:	No, you can't, you can't, no sir

I challenged Marlene in what might seem an unorthodox move, claiming that there is a square root of -1, and that it is -1. I did this because I wanted to see how convinced she was in her thinking. This goes beyond Schifter's skills, but resonates with the idea that teachers can challenge and provoke learners to really think through their ideas. In this case, it was successful, Marlene had the strength to stand up to me and stick with her answer. This shows that she sees that the mathematics makes sense, and is the authority on what is right and wrong, rather than the teacher. In fact, she has a productive disposition (Kilpatrick et al. 2001) towards mathematics, believing that she has the power to make sense of it. I end this section with her conviction.

Marlene:	What number can you square that you get negative one.
Mr Peters:	That's what I'm asking you Marlene, don't ask me, I'm asking the questions
Marlene:	I'm telling you sir, there is no number

Conclusions

This study demonstrates that very few learners in my class were comfortable working with higher-level tasks that asked for justification. This is evident from the 65% of learners who responded similarly to Grace and Rethabile. These learners' difficulty was twofold. They struggled to test and justify the conjecture and they had difficulty with the conceptual understanding of the task. They could not understand $x^2 + 1$ to be a value if x is any real value. Here we encounter the interdependence and interwoven nature of the mathematical strands as argued by Kilpatrick et al. (2001). The few learners who understood the task conceptually and showed some procedural fluency, did manage to make some progress on the task. Of these, only two had provided a complete solution. However, I was able to draw on a partial solution to develop some discussion in the class.

My role as a teacher was central in the development of this discussion. I displayed learners' written work on the board and gave them the opportunity to clarify the mathematical validity of their logic. I made sure that both correct and incorrect responses had to be justified by learners, so that they would not think that I only questioned incorrect responses or reasoning. Marlene showed confidence when I questioned her correct response. She managed to maintain the authority of the mathematics against the authority of the teacher. Other learners were also able to do this at other times during the lessons. My analysis showed that I used Schifter's skills to understand the learners' thinking. I also realized that sometimes I would go into traditional teacher mode and listen for particular ideas or talk for the learners. While it is possible to work against such tendencies, other research has shown that teaching will always be a mix of old and new methods (Brodie 2007a). Teachers can accept this, and know that sometimes their teaching will have to shift. There is no such thing as "pure" teaching, either reform or traditional.

This study has shown that teachers can venture beyond the single strand of procedural fluency. For too long, we as teachers have been stuck in second gear, i.e. trying to master mathematical procedures and have been content to call our actions doing mathematics. This study has offered me a new approach to teaching mathematics through the practices of justification and explanation. Learners need to be challenged to explain and justify conjectures using mathematical reasoning and thinking. To achieve success, we teachers need to equip ourselves with new skills and carefully selected tasks that will assist us in reaching our goals. My study has shown that I can venture beyond traditional mathematics teaching. I hope to continue to move in this direction.

Introduction to Part 3

The teachers' case studies illuminate many of the complexities of working to engage learners' mathematical reasoning. When teachers choose higher-order tasks, learners may struggle to respond appropriately to these and to teachers' moves to try to engage them with the mathematics. Learners can and do internalize teachers' moves designed to get them to think and talk with each other, but this can take time. Teaching in relation to the five strands of proficiency can work, but developing the strands of strategic competence and adaptive reasoning may be more difficult than that of conceptual understanding and procedural fluency. The case studies also suggested that the challenge of teaching mathematical reasoning is more difficult when working with learners who have weaker mathematical knowledge.

Part 3 takes the same data that each of the teachers worked with and looks across the five classrooms for patterns that emerge and that are not visible through the individual case studies. These four chapters address some of the same questions that the teachers addressed in the case studies as well as some new questions:

- How do learners respond to tasks chosen to elicit their mathematical reasoning?
- How can teachers interact with learners around tasks to engage their mathematical reasoning?
- How can we describe learners' contributions and teachers' responses to these in ways that can help us talk about them in more specific ways?
- What kinds of teaching practices, questions, and moves help to encourage and sustain learners' mathematical reasoning?
- What kinds of dilemmas do teachers experience as they teach mathematical reasoning?
- What can teachers do in response to resistance to new ways of teaching?

In Chaps. 8 and 9, I develop a coding scheme to look at learners' contributions to the lessons and teachers' moves in response to these. These provide the beginnings of a language of description for teacher and learner talk in classrooms, particularly classrooms where mathematical reasoning is a focus. I discuss learner contributions in Chap. 8 and teacher moves in Chap. 9. However, learner contributions and teacher moves co-produce each other, and it was very difficult to separate the two chapters from each other. It is therefore important that they be read together.

In Chap. 8, I show that learners produce a range of contributions, describe these, and show that these contributions can be accounted for by the tasks, the learners' knowledge, and the teachers' pedagogies. In Chap. 9, I focus in more detail on the teachers' pedagogies, using a description of teacher moves, as they produce and respond to learner contributions. I suggest a trajectory of emergence and response to learner contributions in classrooms where mathematical reasoning is encouraged. The trajectory suggests that some contributions are easier than others for teachers to recognize and work with, which has implications for teacher education and further research.

In Chap. 10, I focus in detail on two of the teachers, and show how they experienced dilemmas as they worked with learners' mathematical reasoning. The two teachers are Mr. Peters and Mr. Daniels. Mr. Peters' learners had the weakest mathematical knowledge among the five classrooms and, as he described in Chap. 7, struggled to engage with the tasks. Mr. Daniels' learners had stronger mathematical knowledge and managed to support a collaborative discussion, in which important mathematical insights were gained (see Chap. 4). As I will show in Chap. 9, Mr. Peters and Mr. Daniels showed some similarities in their pedagogies – in particular, they both used 'press' moves to develop and sustain conversations among their learners (see also Brodie 2007b). These conversations presented them with a number of teaching dilemmas, which I show are closely related to some of the contradictions inherent in the 'press' move, which is the key to reform pedagogy.

In Chap. 11, I deal with an episode in Mr. Daniels classroom, where the learners strongly resisted his changing pedagogy. I show that different learners have different reasons for resisting reform pedagogy and that resistance can signal engagement with the new pedagogy rather than retreat from it. Such resistance is to be expected and can be managed by teachers.

Part 3 of this book provides different perspectives on the same classrooms that were discussed in Part 2. My analysis across the classrooms was able to illuminate a range of issues that each teacher focusing on her/his classroom alone could not see. I was also able to confirm some of what the teachers did see and comment on. My analysis confirms that "not telling" is not always the best option for teachers, although sometimes it may be (see also Chazan and Ball 1999); that questions are not necessarily more supportive of learners' reasoning than other kinds of teacher moves (see also Brodie 2007a); and that teachers are likely to develop strategic hybrids between traditional and reform practices – "pure" reform practice is unlikely to be found, and to suggest that it is possible may in fact be inappropriate and unhelpful for both teachers and learners to suggest that it is possible (see also Brodie 2007c; Cuban 1993).

Chapter 8
Learner Contributions

The codes that I develop are strongly related to the Initiation-Response-Evaluation/ Feedback (IRE/F) exchange structure discussed in Chap. 1. This was the form of discourse in all five classrooms in the study most of the time, similar to many other mathematics classrooms. The teacher makes an *initiation* move, a learner *responds*, the teacher provides *feedback* or *evaluates* the learner response and then moves on to a new *initiation*. Often, the feedback/evaluation and subsequent initiation moves are combined into one turn, and sometimes the feedback/evaluation is absent or implicit. This gives rise to an extended sequence of initiation-response pairs. An IRE/F exchange pattern allows for a learner contribution at every other turn and a teacher response. As discussed in Chap. 1, when teachers' questions are closed, or funnel learners towards a particular answer, the teachers' combined initiation and evaluation moves leave little space for genuine learner reasoning. However, even within the IRE/F structure, it is possible to see instances where the teacher initiates and responds to learners in different ways, thus allowing for enhanced talk and reasoning by learners.

A description of a range of teachers' response moves is the subject of Chap. 9. In this chapter, I will describe the range of learner contributions in the five classrooms. As I said in the introduction to Part 3, learner contributions and teacher moves are co-produced between teacher and learners. Teacher moves support learner contributions, which in turn support teacher moves. Thus Chaps. 8 and 9 are closely related and need to be read together.

In this chapter I have two main aims. The first is to present the codes that describe the learner contributions in the data and use these to develop a picture for the reader, of the kinds of talking and thinking that learners engaged in across the five classrooms. Using a numerical count of the codes, I show similarities and differences across the classrooms in terms of the learner contributions. The second aim of the chapter is to begin to account for these differences in terms of the key variables discussed in Chap. 2: tasks and learner knowledge. As the analysis progresses, it will become clear that these two variables are not enough to account for the distribution of learner contributions across the classrooms and that the teachers' pedagogies were fundamental in supporting the range of learner contributions in each class.

K. Brodie, *Teaching Mathematical Reasoning in Secondary School Classrooms*,
DOI 10.1007/978-0-387-09742-8_8, © Springer Science+Business Media, LLC 2010

Learner Contributions and Mathematical Reasoning

In Chap. 1 I argued that an important part of teaching mathematical reasoning is to support learners to voice their mathematical thinking and reasoning, nascent or flawed as it might be. However, it is not sufficient for learners to merely express their thinking, as Heaton (2000) illuminates so clearly. She felt lost, unable to make use of the textbook or her own experience to recognize the mathematics in students' contributions, to ascertain whether students' contributions were helpful or not and to work out ways to engage them. Many of her students made contributions that were procedural, or that noted superficial aspects of the mathematics they were talking about. She did not know how to work with these contributions to take them to a deeper mathematical level. Once learners are talking, teachers need to know how to respond appropriately, how to engage learners' thinking and take it forward. This is not an easy task, particularly, when teachers are faced with unusual learner contributions, which they have not heard before or could not have predicted from their knowledge of learners and the task. Heaton argues that knowing what to do with learners' contributions depends on a teacher's mathematical goals and where she wants to take learners mathematically. My argument is that a language for talking about learners' contributions will help teachers and researchers to think about the manner of reality different kinds of learner contributions to mathematical goals and directions.

In traditional mathematics pedagogy, teachers tend to work with the categories of "right" and "wrong". They affirm correct answers and methods and negatively evaluate incorrect ones (Alro and Skovsmose 2004; Boaler 1997; Davis 1997). When teachers do engage with incorrect answers and methods, they usually aim for the production of correct answers, rather than an understanding of why the answers are incorrect and why the learners might be making errors. Moreover, there is always the possibility of a correct response masking a misconception (Nesher 1987), or of learners producing what they think the teacher wants to hear (Bauersfeld 1988). The research on misconceptions and mathematical reasoning suggests that going beyond the notion of error and trying to understand and engage the reasoning behind learners' errors and other contributions is more productive for both teachers and learners (Ball 1993; Chazan 1993; Russell 1999; Sasman et al. 1998; Smith et al. 1993). Many researchers argue that in order to develop mathematical reasoning in classrooms, learners should be asked to justify and elaborate both correct and incorrect answers. A learner's justifications can help her, other learners and the teacher evaluate the contribution, both in terms of the reasoning it presents and in terms of how it takes the conversation forward in the classroom.

When teachers go beyond traditional teaching, and actively elicit and engage learner's ideas in order to develop conceptual links, promote discussion and develop mathematical reasoning, they are often confronted by a range of learner contributions. These contributions might be correct, incorrect or partially correct, well or poorly expressed, relevant or not relevant to the task or discussion, productive or unproductive for further conversation and development of mathematical ideas. Interacting with a range of learner contributions makes teachers' decisions about how to proceed and when and how to evaluate learner thinking very complex. In order to capture the complexity of learner contributions to the lessons, I focused on three important dimensions of their contributions: completeness and correctness of contributions, mathematical reasoning and insights, and learners' grappling with mathematical ideas. These form part of a coding scheme, or language of description, that will be described below.

This work is informed by two important constructivist insights. First, errors that are produced by misconceptions cannot be "treated" by the teacher and replaced with correct ideas; they can only be transformed through the learner reorganizing the ideas that produce them (Smith et al. 1993). Second, errors and misconceptions are a normal part of learning mathematics and cannot be avoided. Describing the errors that learners make does not present a deficit view of learners. Trying to understand how errors emerge in the classroom and how teachers respond to them is a useful way to normalize teacher interaction with learner errors. Since teachers' goals must be to teach appropriate mathematics, they need to orient their thinking and teaching towards learner errors in some way; the aim here is to elaborate how they do this. So it is important to think about errors within a broader range of learner contributions.

Describing Learner Contributions

Six categories can describe almost all the learner contributions in this study. These six are presented in Table 8.1, with an example in response to the task: *can $x^2 + 1$ equal 0 if $x \in R$*. Each contribution is described in detail below.

In developing the above coding scheme, the first important distinction was between responses that could count as complete and correct responses to a task, and those that were partial responses in some way, either incomplete or incorrect. *Complete, correct* responses are responses that provide an adequate answer to a particular task or question. *Partial* responses are those that are either incomplete or incorrect in some way. There are three kinds of partial responses, one of which is incorrect, one of which is incomplete, and one of which is both. An *appropriate error* is an incorrect contribution that mathematics teachers and educators would expect at the particular grade level in relation to the task. Appropriate errors are

Table 8.1 Examples of different kinds of contributions

Contribution	Example
Basic error	$x^2 + 1 = 2x^2$
Appropriate error	If x is a negative number, you can write it as $-x$
Missing information	x^2 is always greater than zero
Partial insight	As you substitute lower numbers, the value of $x^2 + 1$ decreases.
Complete, correct	For $x^2 + 1$ equal to zero, x^2 must be equal to -1. But the square of any number cannot be negative
Beyond task	The square root of -1 squared [$(\sqrt{-1})^2$] equals -1, and then you say $-1 + 1$, then you get 0

distinguished from basic errors, which will be discussed below. A contribution that is *missing information* is correct but incomplete and occurs when a learner presents some of the information required by the task, but not all of it. A contribution that shows *partial insight* is one where a learner is grappling with an important idea, which is not quite complete, nor correct. Although *partial insights* are both incomplete and incorrect, they actually suggest more of a relative presence than the other two partial contributions, in that the learner is offering an insight and is usually grappling with an important mathematical idea.

A second distinction that was important was between appropriate errors and basic errors. *Basic errors* are errors that one would not expect at the particular grade level. *Basic errors* are in a different relation to the task from appropriate errors, because they indicate that the learner is not struggling with the concepts that the task is intended to develop, but rather with other mathematical concepts that are necessary for completing the task, and have been taught in previous years. Finally, contributions that go *beyond task* requirements are contributions that are related to the task or topic of the lesson but go beyond the immediate task and/or make some interesting connections between ideas.

All coding of learner contributions is in relation to a particular task and a particular grade level. For example, what counts as a *basic error* is different in a Grade 3 and a Grade 10 classroom and what counts as a *partial insight* or a *beyond task* contribution will be different in different grades and for different tasks. Almost all of the learner contributions could be categorized into the above categories, with two exceptions. First, a small number of contributions in each class were so unclear that I could not make out the learner's meaning. Second, there were two sets of contributions in Mr. Daniels' lessons in which learners commented on and resisted Mr. Daniels' pedagogy. These are discussed in Chap. 11.

This classification scheme for learner contributions to the lessons is useful in a number of ways. First, it provides a way to characterize what happens in a lesson. A lesson with many complete responses will look very different from one with many partial responses, and both of these will look different from a lesson with many basic errors. Second, the kinds of contributions suggest the affordances and constraints (Greeno and MMAP 1998) that are operating in the lesson. Different kinds of tasks, questions and ways of interacting will enable different kinds of learner contributions, which in turn afford and constrain different kinds of discussions and mathematical reasoning. Third, this classification provides a language for researchers and teachers trying to understand how teachers work with learner contributions to develop mathematical reasoning.

Distribution of Learner Contributions

Table 8.2 shows the distributions of learner contributions across the five classrooms.

The distribution across the categories shows an interesting set of similarities and differences among the learners in the different classes. The two grade 10 classrooms

Table 8.2 Distributions of learner contributions across the classrooms[a]

Teacher	Mr. Peters	Ms. King	Mr. Daniels	Mr. Nkomo	Mr. Mogale
Grade /knowledge	10/weak	10/strong	11/strong	11/weak	11/strong
Basic error	23 (21%)	1 (1%)	0	0	1 (1%)
Appropriate error	21 (19%)	14 (17%)	13 (25%)	11 (25%)	8 (10%)
Missing information	12 (11%)	8 (10%)	6 (11%)	16 (36%)	19 (25%)
Partial insight	9 (8%)	2 (3%)	16 (30%)	1 (2%)	13 (17%)
Complete Correct	38 (35%)	48 (59%)	10 (19%)	16 (36%)	36 (47%)
Beyond task	3 (3%)	7 (9%)	6 (11%)	0	0
Other	4	2	2	0	0
Total number of contributions	110	82	53	44	77

[a]The unit of coding for learner contributions was slightly larger than the turn. Sometimes it would take a number of turns for a learner to state an idea that could count as a contribution. Every idea that was stated by a learner was counted and elaborations on previous ideas were counted as new contributions.

have a similar distribution across most of the categories except in *basic errors* and *complete, correct contributions* where there is a substantial difference. This difference can be explained by the differences in the learner knowledge in the two classrooms as well as the teachers' pedagogies, as will be discussed below.

The three Grade 11 teachers show an interesting set of similarities and differences, in relation to each other and to the Grade 10 teachers. If we take the three kinds of partial responses together, we see that there were in total 66% and 64% respectively in Mr. Daniels and Mr Nkomo's lessons, 51% in Mr. Mogale's lessons and 39% and 30% respectively in Mr. Peters' and Ms. King's lessons. So in this respect, Mr. Daniels' and Mr. Nkomo's learners are similar to each other and very different from the Grade 10 learners, with Mr. Mogale's learners somewhere in between. These similarities and differences can be partially accounted for by the nature of the tasks they were using. The Grade 11 tasks required learners to compare and contrast different graphs as they shifted horizontally and vertically, and to infer general characteristics of the shifts inductively. The tasks asked for observations and asked questions such as "what did you notice" (see Chap. 2 and Appendix for more detail). These kinds of task demands tended to produce *partial responses*, because learners did not necessarily deal with all of the information in the task. They often wrote or reported contributions with *missing information* and as they made inferences based on limited information they made *appropriate errors*.

Looking within the *partial response* category we find that it shows differences among the Grade 11 teachers within the same task constraints. Mr. Daniels' and Mr. Nkomo's learners were similar with respect to *appropriate errors*, while Mr. Mogale's learners were different. Mr. Mogale's and Mr. Nkomo's learners were somewhat similar in relation to *missing information* contributions, and Mr. Daniels' learners were different. The three sets of learners were very different from each other in relation to *partial insights*. To account for these similarities and differences, the qualitative analysis below suggests that three variables: tasks, learner knowledge, and teacher pedagogy are important.

Another important difference is the low number of *complete, correct responses* among Mr. Daniels' learners in relation to all of the other teachers. Given that their knowledge was stronger than both Mr. Nkomo's and Mr. Peters' learners, this is a surprising finding. Since the distributions are in percentages, one obvious reason for fewer *complete, correct* contributions is that there were relatively more of the other kinds of contributions in Mr. Daniels' lessons, particularly *partial insights* and *beyond task* contributions.[1] The substantive reason for the distribution among these three kinds of contributions is Mr. Daniels' pedagogy, as will be argued below and in Chap. 9.

The above is a brief, first-level description of learner contributions in the lessons, suggesting that tasks, learner knowledge, and teacher pedagogy account, at least partially, for different distributions of learner contributions across the classrooms. To understand more fully how teachers support and engage these contributions requires a more in depth analysis of the different contributions and how they emerged in the classrooms.

Accounting for Learner Contributions

In order to account for the distributions of learner contributions, the discussions of learner knowledge and tasks in Chap. 2 is important. Table 8.3 should remind the reader of these dimensions of the classrooms in the study.

The Grade 11 teachers all used the same tasks, which were inductive and which supported procedures with connections to meaning. Mr. Daniels' and Mr. Mogale's learners had strong mathematical knowledge while Mr. Nkomo's learners had weak mathematical knowledge The two Grade 10 teachers used similar tasks, which were mainly deductive and which varied from "doing mathematics", through procedures with connections to procedures without connections. Ms. King's learners had very strong mathematical knowledge while Mr. Peters' learners had weak mathematical knowledge. I now turn to a discussion of the different kinds of learner contributions.

Table 8.3 Variation across teachers in tasks, learner knowledge and SES

	Learner knowledge	Stronger	Weaker
tasks			
Grade 11 Inductive Procedures with connections		Mr. Daniels Mr Mogale	Mr. Nkomo
Grade 10 Deductive (with some inductive) Procedures with and without connections, doing mathematics		Ms. King	Mr. Peters

[1] This also raises an interesting question: what is an appropriate range of complete, correct contributions in lessons where teachers are exploring learners' reasoning.

Basic Errors

Basic errors were substantially present only in Mr. Peters' classroom and accounted for 20% of the learner contributions in his lessons. These errors occurred throughout the lessons, were for the most part noticed and taken up by Mr. Peters and significantly influenced the flow of the lessons. These errors tended to cluster together in relation to a particular idea or task and can be classified into three main groups, which will be discussed below. Mr. Peters dealt with basic errors by trying to find out the exact nature of the error and then working to correct it.

The first set of *basic errors* related to algebraic manipulation of the expression $x^2 + 1$, in the task: *Consider the following conjecture: "$x^2 + 1$ can never be zero". Prove whether this statement is true or false if $x \in R$.*[2] During discussion of this task and in their written work, a number of learners claimed that: "x has unknown value, so it can be taken to be 1", "x is equal to 1 because there is a 1 in front of the x"; and "because there is a 1 in front of the x, $1x^2 + 1 = 2x^2$".

A second set of *basic errors* occurred when learners tried to substitute fractions into the expression $x^2 + 1$ in order to see whether it could equal zero. Only a few learners had done this in their written work and Mr. Peters introduced it to the class as a possibility and suggested that they try substituting ½, asking "what's ½ times ½". Learners responded with a range of answers including ½, ¼ and 1. In trying to deal with this error, Mr. Peters had to use more fractions, which brought up more errors. For example, Mr. Peters showed that 1/2 times 6 can be written as ½+½+½+½+½+½; some learners suggested that this gives 6/12, while others suggested 1/12.

A third set of *basic errors* in Mr. Peters' class related to substituting numbers in algebraic expressions. As learner substituted values into expressions such as $x^2 + 1$, $-2x$, $3x^2$ and others, it became clear that many learners had trouble with actually substituting values for x. For example, Lebo said that if $x = -1$, x^2 gives -1 gives -1 which is -1, Samantha said if $x = -1$, $3x^2$ is -3, Ahmed said that if $x = -9$, he got $-9 + x^2$ when substituting into $x^2 + 1$ and a number of learners said that if $x = -3$, $3x^2 = 9$, -9 and $= -3x^2$.

The fact that *basic errors* occurred only in Mr. Peters' classroom can be accounted for in two ways. First it is clear from the learner interviews discussed in Chap. 2 that his learners had very weak mathematical knowledge. However, Mr. Nkomo's learners also had weak mathematical knowledge and no basic errors surfaced in his lessons. An analysis of Mr. Peters' pedagogy, in particular his way of dealing with both *appropriate and basic errors* suggests that his pedagogy was key in allowing basic errors to surface.

Mr Peters spent a lot of time asking learners to justify their answers (see also Chap. 7). His usual response to a *basic error* was to ask the learner to explain how s/he got the answer. As the learner explained and others responded, other *basic errors* emerged, as shown above (see also Chap. 9). Mr. Peters spent a lot of time discussing *appropriate errors* with learners, again providing opportunities for more

---

[2] A detailed discussion of the range of responses to this task can be found in Chap. 7.

errors to appear. Mr. Peters allowed errors to remain on the agenda in his lessons for a number of reasons: to find out how widespread the errors were in the class; to understand why learners were making these mistakes; to help learners understand why they were incorrect; and to put himself in a position to correct them. He also tried to find out which learners did not share the errors and might help him to correct them. In order to do this, Mr. Peters often called on other learners to explain why particular errors were incorrect. This move often produced yet more errors. So the basic errors in Mr. Peters' lessons can be accounted for both by the weaker knowledge of his learners and his pedagogy of focusing on errors.

Appropriate Errors

An *appropriate error* is an incorrect contribution that might be expected as learners grapple with new ideas in relation to the task. *Appropriate errors* may indicate misconceptions. Table 8.2 shows that there were a significant number of these in all the classrooms: 25% in Mr. Daniels' and Mr. Nkomo's lessons; 19% in Mr. Peters' lessons; 17% in Ms. King's lessons; and 10% in Mr. Mogale's lessons. In this section, I show that the *appropriate errors* are predominantly related to the tasks and learner knowledge.

In the Grade 11 classrooms, there were two main clusters of *appropriate errors*. The first was in activity 1, when learners had to say what happens to points on the graph $y=x^2$ as it shifts 3 units to the right or 4 units to the left. Learners made comments like the following: "all your coordinates sit in the first quadrant" (referring to the shift to the right) and "all your coordinates sit in the second quadrant" (referring to the shift to the left); "the new graph does not cut the y-axis"; "there are no y-intercepts for either shift" and "negative points become positive points". These contributions suggest a conception of a bounded graph, and that the learners are attuned to the image of the graph on their page at the expense of the function that produces the graph. When learners claim that all the coordinates sit in the first quadrant, they are not seeing that although the graph has shifted, it can still be extended into the second quadrant and in fact the two graphs have exactly the same domain and range. The difference is in the relationship between the x and y-values on each graph. In all three classrooms, these contributions were made during report backs from previous group work, suggesting that the issue was not resolved in the groups.

A second set of *appropriate errors* occurred in the relationship between the sign in the bracket of $y=(x-p)^2$ and the sign of the x-coordinate of the turning point $(p;0)$. In all three Grade 11 classrooms, learners confused the sign of p when p is negative, for example they identified the value of p in $y=(x+4)^2$ as 4 and so did not understand why the turning point was $(-4;0)$. In Mr. Nkomo's and Mr. Daniel's lessons, learners tried to deal with the difference in sign by suggestion that "if you take that positive four out, you put it on the y side, it will turn to a negative", an inappropriate manipulation of the equation. This did not happen in Mr. Mogale's lessons, for two possible reasons. First, because Mr. Mogale's learners' knowledge was stronger, they may not have considered the incorrect manipulation of the equation as an option. Second, Mr. Mogale focused his questions on how the axes of symmetry, and hence the turning points, shift as the graphs shift. So Mr. Mogale focused his learners on the graphs while the learners in the other two classrooms focused predominantly on the equation.

The appropriate errors in Mr. Peters' lessons also clustered around particular tasks and concepts. The first, which was widespread (see Chap. 7) was learners' insistence that x^2+1 could not be simplified, and therefore could not equal zero. As Grace and Rethabile put it: "you can never get the zero because it can't be, because they unlike terms, you can only get, the answers only gonna be x squared plus one, that's the only thing that we saw because there's no other answer or anything else". Mr. Peters spent some time discussing this appropriate error with the class and realized that learners were not seeing x^2+1 as a variable expression that could take on a range of values depending on the value of x; rather they were focusing on the syntactic elements of the expression – two terms separated by an addition sign. A related, second set of *appropriate errors* occurred when the learners claimed that an expression like $-x$ or $-2x$ is negative because of the negative sign, showing very clearly how they focus on the syntactic elements of the expression rather than its meaning.

Mr. Peters therefore developed a new set of tasks and spent most of his lessons trying to help learners to generate meaning for algebraic expressions.

State whether the following expressions in terms of x are:

(i) Always ≥ 0
(ii) Always ≤ 0
(iii) Sometimes positive, sometimes zero, sometimes negative, depending on the value of x.

    a) x
    b) $-2x$
    c) x^2
    d) $3x^2$
    e) $-x^2$
    f) $(x+1)^2$
    g) $-(x+2)^2$
    h) $2(x-3)^2$

As learners worked through these tasks, a third *appropriate error* arose, in relation to whether $3x^2$ is a perfect square. Some learners argued that it is, because they understood $3x^2$ to mean $3x$ times $3x$. Other learners disagreed, saying that since 3 is not a perfect square, $3x^2$ is not a perfect square. This is a subtle issue relating to the order of operations in algebraic expressions.

There were similar *appropriate errors* in Ms. King's lessons, particularly in relation to the meaning of x and x^2. Some learners argued that $-(1)^2$, rather than $(-1)^2$ is an appropriate instantiation of x^2. They did not understand that in this context the square applies to the whole of *x*, which is −1, rather than to the 1 only. This is also an issue of order of operations, similar to the $3x^2$ in Mr. Peters' class. Also, similar to those of Mr. Peters' learners, some of Ms. Kings' learners claimed that −*x* represented a negative number. A major difference between the two classrooms was that these errors were widespread among Mr. Peters' learners while very few learners in Ms King's class made them, because of their stronger knowledge. Ms. King was able to draw on other learners for correct responses and explanations, which came very quickly and were usually accepted by the learner who had made the error.

In this section, I have shown that there were clear task-related similarities in *appropriate errors* across the classrooms and that these arose as learners brought their current understandings to the tasks. There were fewer *appropriate errors* in Mr. Mogale's and Ms. King's lessons because of the stronger knowledge of their learners. Teacher pedagogy was not an important variable in producing *appropriate errors*. Moreover, the teachers' responses to *appropriate errors* were complex and relate closely to their responses to the other partial contributions. For these reasons, the teachers' responses are not discussed here but in Chap. 9.

Missing Information

A contribution that is *missing information*, goes some way to answering the question, presents some correct information, but is incomplete. Not everything required is presented, but there are no errors. In some cases, a number of these responses follow each other, and together the responses create a complete response. There were somewhat more of these in Mr. Nkomo's and Mr. Mogale's lessons (36% and 25%) than in the other three classrooms (11%, 10% and 11%).

Most of the *missing information* contributions in the Grade 11 classrooms occurred during report backs, together with *appropriate errors*, as discussed above. As discussed in Chap. 2, the Grade 11 tasks were exploratory, asking learners to explore shifting graphs and to "write what you notice". These kinds of task demands, tended to produce *missing information* responses because learners could choose what they noticed and may not have discussed all possibilities. The following is an example from Mr. Daniels' class:

> Okay, what our group basically came up with, is that um, if we move the graph along the x-axis, to wherever, the x-values are gonna change and the y-values are gonna remain constant, (*slides a transparency over an original graph on the OHP to show changing values*) and if you move it along the y-axis, then uh, then the y-values change accordingly and the x-values remain constant, and the coordinates are also gonna change

The above contribution correctly argues that the *x* and *y* values of corresponding points on the graph will change when the graph is moved horizontally and vertically. It leaves out important information about the amount of change, which was asked for in the question. Similar missing information contributions occurred in the three Grade 11 classes but Mr. Daniels and Mr. Nkomo dealt with them differently from Mr. Mogale. Mr Daniels and Mr. Nkomo often called on other groups to add to what had been said, so for example after interacting with the group who made the above comment, Mr. Daniels called on a second group who reported back:

> This is for the first one, you move it three places, three units to the right, then the co-ordinates have all changed and all your x-values have increased by three. Um, then if you look at the table, if you fill in the table, then there will be a pattern that will go, one unit at a time will go nought, one, two, three, four, five, six. There's, then there'll be a new equation – it will be, y equals x squared plus three, and then for one point two …

Although this contribution still has *missing information*, and an *appropriate error*, this group added additional information not provided by the first group, that the x-values increase by 3. I show in Chap. 9 how both Mr. Daniels and Mr. Nkomo focused on missing information contributions, sometimes at the expense of appropriate errors, encouraging learners to develop more complete contributions. Mr Mogale encouraged learner–learner interaction, and because his learners had strong knowledge, they often completed *missing information* contributions and corrected their classmates' *appropriate errors*.

In Mr. Peters' class, *missing information* contributions almost always related to learners' not seeing all possibilities for an algebraic expression. When Mr. Peters asked "what does x mean to you?" a learner contribution was: "x has unknown value". When he asked for an example, learners gave only positive numbers and zero. So Mr. Peters asked again, and got negative numbers. Mr. Peters built these responses to a complete response, through his questioning of learners in IRE patterns. Ms King also built from incomplete to complete responses using closed questions and IRE patterns. For example, when looking at the expression $f(a+h)$, she asked what is $a+h$, and received the responses, "a variable or a subject"; "an element", and "an element of the real numbers". She asked for more information, but when a more complete response was not forthcoming, told the learners to look back at their worksheets for a definition, which was "its an element of Set A" and which seemed to be the complete response she was looking for.

So *missing information* contributions are related both to the nature of the tasks and to teacher pedagogy. They are related to the nature of the tasks in that, inductive exploratory tasks are more likely to produce missing information contributions. They were related to the teachers' pedagogy in that, report backs from group work tended to produce missing information contributions and because all of the teachers worked to complete these, although in slightly different ways. The teachers' responses to these contributions were closely related to their responses to *appropriate errors* and are discussed in detail in Chap. 9.

Partial Insights

Partial insights are contributions that present an important idea, not quite fully formed but which suggest that the learner is grappling with some important aspects of the tasks or concepts, and making connections between ideas. These occurred mainly in Mr. Daniels' lessons (30%), with some in Mr. Mogale's lessons (17%), a few in Mr. Peters' lessons (8%) and almost none in Mr. Nkomo's and Ms. King's lessons (2% and 3%).

In Mr. Daniels' and Mr. Mogale's lessons *partial insights* occurred as the teacher or other learners pushed a learner to clarify or elaborate their thinking. An example from Mr. Daniels' classroom occurred when Michelle asked why the signs of the turning points [(3;0) and (−4;0)] of the graphs $y=(x-3)^2$ and $y=(x+4)^2$ are the opposite of the signs in the brackets. Michelle's contribution generated a discussion in which many *partial insights* came up, as learners tried to make sense of each other's ideas and respond to them in order to answer Michelle's question. One of these was Winile's argument that:

the +4 is not like the x, um, the x, like, the number, you know the x *(showing x-axis with hand)*, it's not the x, it's another number … you substitute this with a number, isn't it, like you go, whatever, then it gives you an answer

and later

we're not supposed to get what x is equal to, we getting what y is equal to, so we supposed to, supposed to substitute x to get y

Winile's insight was that you could not make a direct link from the equation to the graph as Michelle was trying to do, because you had to operate on the equation first. She had not yet seen that there is a more direct relationship than she was imagining, but her contributions focused on the underlying relationships between the equation and graph rather than on the surface features of each and functioned to focus other learners on these as well (see Chap. 4).

Partial insights occurred as Mr. Daniels and Mr. Mogale encouraged learners to speak to each other, rather than to them. They both avoided making suggestions except in cases where they thought their suggestions would refocus learners on to the important points of the discussion. Enabling learner–learner talk was an explicit goal of both of their pedagogies and they both spoke often about how they tried to achieve this. Both also worked hard to communicate norms of communication to their learners.

It can be argued that *partial insights* emerged in these classrooms because of the stronger learners, and this might be part of the reason. However, we should note that there were very few *partial insights* in Ms. King's lessons' and there were some in Mr. Peters' lessons, even though he had the weakest learners. Mr. Peters also managed to generate some discussion in his lessons, as will be discussed below and in Chap. 9, which led in some cases to *partial insights* even though his learners had weak knowledge. The distribution of *partial insights* across grades and learner knowledge, together with analyses of the teachers' pedagogies suggests that pedagogy is crucial in supporting *partial insights*. This will be discussed in more detail in Chap. 9.

Complete, Correct Contributions

These were distributed differently across the classrooms: 59% of the total learner contributions in Ms. King's lessons; 47% in Mr. Mogale's lessons; 35% in

Mr. Peters' classroom; 36% in Mr. Nkomo's classroom, and 18% in Mr. Daniels' classroom, suggesting, as with *partial insights*, that they are not easy to relate to the tasks or learner knowledge. There were two kinds of *complete correct* contributions: relatively short, quick responses to relatively closed questions; and longer, more substantial responses to more open tasks and questions. The teachers dealt differently with these.

Longer complete contributions tended to occur during report backs from the group work. In all five classrooms, some of the groups had complete, correct responses and the teachers used these strategically, usually after discussion of some of the partial contributions. All of the teachers pressed further on these contributions, suggesting that they were not endpoints in themselves, but the sites for further discussion. They all wanted these contributions to be visible and noted by other learners as complete and correct and some of the work that they did served this purpose. A range of other teacher purposes were: to show that even when a contribution was complete and correct, more could be gained from it: to show different strategies; to elicit some deeper thought as to what counts as an explanation; and to encourage learners to stand up for their responses and defend them from challenges, whether from other learners or the teacher. Taking complete, correct contributions further, shows that all of these teachers were working beyond the "traditional" approach of only affirming correct responses.

The teachers dealt differently with the shorter *complete, correct* contributions. These were usually short quick responses to relatively closed questions. Teachers most often affirmed these contributions. These kinds of questions, responses and affirmations are most often associated with the IRE exchange structure and traditional teaching, which was the case in this sample. However, my examples below suggest that even in this relatively uniform practice there are still a variety of functions and consequences.

In Mr. Nkomo's lessons, many of the shorter complete responses came towards the end of lessons, as he both summarized what had happened in the lessons and as he developed the more general relationships between the equation $y = a(x-p)^2 + q$ and the particular graphs. Ms. King developed a worksheet with many closed questions (see Chap. 2), for example, if $g(x) = 2x^2 + 3x - 1$, what is $g(1)$, $g(-2)$, $g(a)$, $g(ab)$, $g(a+b)$. Learners wrote their answers on the board and Ms. King and the class evaluated them and corrected them where necessary. Most of the learners managed these and so there were many *complete, correct* responses, evaluated affirmatively. Also, as she taught, Ms. King would develop important ideas through sequences of questions that obtained *complete, correct* responses (see Brodie 2006 for more detail). In Mr. Peters' lessons, many of the shorter complete responses occurred as he corrected *basic errors*. For these three teachers, the IRE functioned traditionally to achieve *complete, correct* responses.

Both Mr. Mogale and Mr. Daniels used IRE sequences similar to those used by the other three teachers. In addition, there are instances where these question and answer sequences played a different role. For example, in Mr. Daniels' class, when the learners were discussing the fact that the equation $y = (x+4)^2$ had turning point $(-4;0)$, they spoke about the turning point being -4 rather than $(-4;0)$. This is a perfectly appropriate way of speaking about the turning point, which mathematics teachers and mathematicians use. Although we know that the point has two coordinates, we know that the y-value of the turning point is zero, and so we only speak about the x-value. In this case though, Mr. Daniels raised the issue with the class, in an IRE sequence.

Mr. Daniels:	Okay, now my question is, what is the turning point there really
Learner:	It's minus four
Mr. Daniels:	Okay now, what is a point, a point is made up of what
Learner:	x- and y-coordinates
Mr. Daniels:	x- and y-coordinates, good

In the context of a longer discussion with many *partial insights* and with many twists and turns, the IRE sequence above and the closed question: "A point is made up of what" reminded the learners of what they knew but were not using to think about the task, that the turning point has two co-ordinates. This supported them to think about the relationship between the two co-ordinates and hence the relationships between the turning point and the equation in new, helpful ways, which helped them to make progress with the discussion (see Brodie 2007c for more detail).

So, there are two kinds of complete, correct contributions in these lessons, both of which play important roles in the lessons, and which are dealt with by the teachers in interesting ways. The shorter complete responses occur across the teachers, irrespective of tasks or learner knowledge. They come up when teachers use the traditional IRE sequence in traditional ways, to teach or summarize ideas, or correct errors relatively quickly and efficiently. There is also evidence that the IRE can be used in less traditional ways: in order to make important points relatively quickly and efficiently in order to enable the discussion to continue. The longer *complete, correct* responses also occurred across the classrooms. In these cases, all the teachers pressed further on these responses, showing that correct responses are not only valuable because they are correct, but because in thinking more about them, they can support further insights. So *complete, correct* responses are most strongly related to teacher pedagogy and also to learner knowledge.

Going Beyond the Task

In three of the five classrooms, learners made contributions that went beyond the task demands, making links between ideas in interesting ways and opening up potentially interesting avenues for conversation. There were relatively few of these in the three classrooms: 3 (3%), in Mr. Peters', 6 (11%) in Mr. Daniels', and 7 (9%) in Ms. King's. The three teachers took these up in ways that gave them more significance as opportunities for learning than the numbers suggest.

In Ms. King's class there were two sets of *beyond task* contributions. The first came up when Ms. King pressed further on a *complete correct* response to the question of whether $x^2 + 1$ could equal zero, asking the class whether the learner had explained why the square of any number can never be negative and whether he should have.

The latter question took the learners beyond the task, requiring them to think about when it is necessary to articulate an explanation. Some learners indicated that they thought he should explain why, while others indicated that there was no need for an explanation because "its common knowledge", i.e. explanations are not always necessary if there is an assumed shared base of understanding. Ms. King allowed some discussion of this issue and did not take a final position herself, leaving the question open.

The second set of *beyond task* contributions in Ms. King's class came up when some learners took the problem to another mathematics teacher who told them that $x^2 + 1$ can be zero if they considered complex numbers, and i as $\sqrt{-1}$. This group of learners discussed this with Ms. King prior to the lesson and when she had finished with the responses from the different groups, she asked them to explain what they had found. The discussion led to some *appropriate errors* as discussed above.

Ms. Kings' responses to *beyond task* contributions were that she noticed them, tried to press for further contributions from learners and tried to build on their ideas. When this proved difficult, she took the opportunity to explain some mathematics that came out of their ideas. She provided clear, well-structured and elaborate explanations of mathematical concepts that responded to the learners' contribution and took them further.

In Mr. Peters' lesson there were three *beyond task* contributions, all made by the same learner, Lebo. First Lebo asked why −2 times 0 gives positive 0 because, as Mr. Peters had recently emphasized, multiplying a positive by a negative number should give a negative number. Lebo argued that the answer to −2 times 0 should be −0. Lebo's contribution began a conversation that continued for about 70 turns and included *appropriate errors, missing information and partial insights*. During the conversation, six learners, including Lebo made individual contributions, taking between 1 and 12 turns each. Some examples of contributions were: "zero is neutral", "zero is neither negative nor positive because of where it is on the numberline", "positive zero and negative zero is the same", "zero is like x because sometimes its positive and sometimes its negative", "zero is nothing", "zero is not nothing because we get problems with zero in like zero plus one and zero times zero and, "zero is a number because its on the numberline". The teacher responded after each learner turn in a number of different ways. Sometimes, he pressed the learner to elaborate her/his idea, sometimes he called on another learner to contribute, sometimes he repeated a contribution, and sometimes he inserted his own contributions. He often refocused learners on the question in order to keep the point of the discussion clear, for example he asked: "Does it make a difference if we write positive zero or negative zero"; and "Why do we write it just as zero, why don't I write negative zero?"

A detailed analysis of the conversation shows that there were many places where learners built on and challenged each other's contributions and reflected on their own contributions. Lebo returned to her question a number of times, indicating that it had not been answered. Mr. Peters ended the conversation by saying they would agree to think of zero as neutral. Lebo's facial expression indicated that she was not happy with how Mr. Peters ended the discussion. Mr. Peters acknowledged her dissatisfaction, which led to another interesting interaction and another *beyond task* contribution:

Lebo:	Sir, can I please ask you, if they say you are nothing, then, and, you have something that they, that they don't have, Sir
Mr. Peters:	Ja, ja they can't, if someone says you are nothing, they can't say, you are say, you a positive nothing
Lebo:	Yeah, right
Mr. Peters:	They can't say, you are a negative nothing, they can't say so, but, if he says, hey you a two, hey, you a positive two, can you see, he's putting value there, you see, you see, but if he says you are a negative two, then it means that something's bad, you see, that's in the connotation of negative, but if he says you are nothing, are you positive nothing, there's no difference if I say you are positive nothing or you are nothing, it's exactly the same. Lebo, you are a positive two

Lebo did not go back to the original context of her question. Instead she made the link with everyday talk where people call each other "nothing". From the tone of her voice, and her comment "you have something that they don't have", it seems like she has been a victim of such talk and is trying to find ways of dealing with it. In this way, she diverted attention from her original question. Mr. Peters tried to take up this issue, in a way that affirmed her feelings and her thinking, saying "Lebo you are a positive 2". He drew on similarities between the everyday use and the mathematical use of nothing, but did not emphasize the differences. In this way, he again made his point that zero is neither positive nor negative, but did not answer Lebo's original question.

So the few *beyond task* contributions can be somewhat accounted for by learner knowledge and teacher pedagogy. However, this cannot be claimed very strongly, given that very few of these were seen in only three classrooms.

Summary

In accounting for the distribution of learner contributions across the lessons, I have argued that the key variables are learner knowledge, tasks and pedagogy. I have looked at how these vary across the teachers as well as the ways in which they are related to the distributions of learner contributions. Table 8.4 shows the relationships as discussed in the chapter, which are summarized below

Basic errors occurred in significant numbers in only one classroom (Mr. Peters') but in that classroom they were clearly related to the learners' weak mathematical

Table 8.4 Key variables and learner contributions

	Learner knowledge	Tasks	Pedagogy
Basic errors	✓		✓
Appropriate errors	✓	✓	✓
Missing information		✓	✓
Partial insights		✓	✓
Complete and correct		✓	✓
Beyond task	✓		✓

knowledge. However, given the large number of *basic errors* in Mr. Peters' class-room (20%), and the very small number (or none) in the other classrooms, I also argued that *basic errors* could also be accounted for by Mr. Peters' pedagogy in that he noticed and worked with both *appropriate* and *basic errors* extensively, which increased their visibility and allowed for additional errors to enter the conversation.

Appropriate errors occurred in all the classrooms and were distributed differently across the classrooms. I argued that these could be accounted for predominantly by learners' knowledge and the tasks. It was clear that the kinds of *appropriate errors* in each classroom were task-related and it was also the case that particular *appropriate errors* were more widespread in Mr. Peters' class than in Ms. King's class. Some elements of Mr. Mogale's pedagogy accounted for fewer *appropriate errors* in his class.

Missing information contributions were task-related in that, the tasks that asked learners to compare and contrast produced more of these, predominantly in Mr. Nkomo's classroom. *Missing information* contributions occurred mainly during report-back sessions, and so can be related to pedagogy in that sense. These contri-butions could not be related to the variations in learner knowledge across the class-rooms. *Partial insights* were most clearly related to pedagogy in that they occurred in the three classrooms where the teachers managed to support learner–learner discussion, Mr. Peters', Mr. Daniels' and Mr. Mogale's lessons (see Chap. 9 for more detail). *Partial insights* can also be argued to be task-related in that higher-level tasks are necessary, although not sufficient to enable learner engagement.

Complete, correct contributions were of two kinds: extended and short. *Extended complete, correct* contributions were mainly task-related in that the tasks demanded these kinds of contributions, either through the compare and contrast responses or through justification. Similar to *missing information* contributions, *extended complete correct* contributions were related to the tasks and pedagogy similarly across the teachers, in that they occurred during the report back sessions. *Short, complete correct* contributions were related to pedagogy, again similarly across the teachers, in that they were produced by relatively closed and constrained questions and in more traditional-looking IRE exchanges.

Finally, *beyond task* contributions were the most complex to account for, particularly given that very few occurred. However, these are significant, and are worth understanding in more detail if we want to try to promote them more broadly in classrooms. I argued that these could be accounted for by both learner knowledge and pedagogy.

In this analysis, I have shown how and why the distributions of learner contribu-tions are created in the classrooms. I have shown that the kinds of contributions depends on tasks, learner knowledge and teacher pedagogy. Given the small and purposeful sample, these results cannot be generalized empirically to other teachers. However, the explication of the relationships between tasks, learner knowledge and pedagogy are generative in understanding how teachers in general might work with learners' thinking.

In relation to tasks, the teachers chose to work with higher-level tasks and I have argued, that these were necessary but not sufficient to enable *partial insights*.

The fact that the tasks required extended responses enabled extended *complete correct* contributions. The particular content of the tasks generated task-related *appropriate errors*. Finally, inductive, compare and contrast tasks generated *missing information* contributions, because they required learners to say what they noticed, rather than to provide a deductive justification.

In relation to learner knowledge, I have shown that weaker learner knowledge produces many more *basic and appropriate errors*, even though the *appropriate errors* are of the same kind among weaker and stronger learners because of a strong association with the tasks. The case of Mr. Peters suggests that teachers of learners with weaker knowledge may be faced with additional challenges in working with learners' mathematical thinking. As Mr. Peters began to work with learners' ideas and contributions, he faced a barrage of errors, which were difficult to deal with. The cases of Ms. King, Mr. Mogale and Mr. Daniels suggests that stronger learner knowledge helps in producing *beyond task* contributions and *partial insights*, but pedagogy is crucially important, both in producing these and in taking them forward in productive ways.

Finally, I have argued that teachers' pedagogy is a crucial variable in producing different kinds of learner contributions. The particular form of report-backs from group work helped, together with the tasks, to produce *missing information* and extended *complete, correct* contributions. Particular teacher pedagogies supported all of the contributions in different ways for the different teachers. In the next chapter I look in more detail at teacher pedagogy, as seen through a description of "teacher moves", showing how these supported and took forward the different learner contributions.

Chapter 9
Teacher Responses to Learner Contributions

In chapter 8 I argued that learner contributions in the lessons I observed were pro-duced by an interaction between tasks, learner-knowledge and teacher pedagogy. Teacher pedagogy can be seen in broad terms, such as the selection and modifica-tion of tasks and the use of group work and report-backs. These aspects of the teachers' pedagogies were important in co-producing the different learner contribu-tions. However, my descriptions in the previous chapter also suggested that a more fine-grained description of teachers' pedagogies, in particular, their moves in response to learner contributions, is necessary. In this chapter I deepen my descrip-tion of the teachers' responses to the different learner contributions by using a cod-ing scheme for the teacher moves in the lessons. As with learner contributions, this coding scheme builds on the IRE/F exchange structure and, drawing on a range of other literature on teacher moves and practices, elaborates the teacher's turns in this structure, the initiation and response turns.

Teacher Moves

My codes describe the function of teacher utterances as they initiate and evaluate. As Mehan (1979) argues, initiation and evaluation are often combined into one teacher turn and so they are fused in form, although not in function. Therefore, my unit of coding was the teacher turn, or sub-turn, when the teacher made more than one move in a turn. A teacher move is constituted by all or part of a teacher turn, which can be described with one code. Thus codes help to determine moves, it is not possible to identify moves prior to coding.

When looking at teachers' responses to learner contributions, a key code is *follow up*, which describes all teacher moves that respond to learner contributions. A teacher move is coded as *follow up* when the teacher picks up on a contribution made by a learner, either immediately preceding or some time earlier. The teacher could ask for clarification or elaboration, ask a question or challenge the learner. Usually, there is an explicit reference to the idea, but there does not have to be. Usually, the idea is in the public space, but it does not have to be; for example when a teacher asks a learner to share an idea that she saw previously in the learner's work.

K. Brodie, *Teaching Mathematical Reasoning in Secondary School Classrooms*,
DOI 10.1007/978-0-387-09742-8_9, © Springer Science+Business Media, LLC 2010

Repeating a contribution counts as *follow up* if it functions to solicit further contributions in relation to the learner's contribution. If a teacher repeats a contribution to affirm it and the discussion ends, then the move is coded *affirm*, not *follow up*.

My "follow up" code is closely to related to Nystrand et al's. (1997) notion of "uptake." However, it is broader than their notion and includes some teacher moves that they might not include. This is discussed in more detail below. An initial coding of my data showed that there were a large number of *follow up* moves which functioned differently, so I further divided this category into different kinds of *follow up*. The five subcategories of follow up are described in Table 9.1.

These codes are informed by various concepts in the literature. *Elicit* is closest to Edwards and Mercer's (1987) "repeated questions imply wrong answers" or Bauersfeld's (1980) "funneling," where the teacher increasingly narrows her questions to help the learner provide the expected answer. This corresponds to a lowering of the task demands (Stein et al. 1996; Stein et al. 2000) and suggests that the learners may produce an answer that they do not own or fully understand. So *elicit* moves can constrain as much as enable learner thinking. It is likely that Nystrand et al. (1997) would not have included *elicit* moves in their notion of uptake because they may not represent a serious consideration of learner ideas.[1] I chose to include *elicit* moves under *follow up*, because for my purposes it is illuminating to distinguish between different kinds of *follow up* rather than to exclude a range of moves that teachers might intend as a *follow up* from this category. *Press* is similar to Wood's (1994) "focusing" and to Boaler and Brodie's (2004) category "probing." *Press* moves include Nystrand et al.'s "authentic" questions (1997), but is not limited to them. Authentic questions are questions to which the teacher genuinely does not know the answer. A teacher might *press* when she does know the answer but wants to give the learner a chance to articulate and hence deepen her thinking, and/or wants to make

Table 9.1 Subcategories of follow up

Insert	The teacher adds something in response to the learner's contribution. She can elaborate on it, correct it, answer a question, suggest something, make a link etc
Elicit	While following up on a contribution, the teacher tries to elicit something new from the learner or other learners. She elicits additional information or a new but related idea to take the lesson forward. Elicit moves often, but not always, narrow the contributions in the same way as funneling
Press	The teacher pushes or probes the learner for more on her/his idea, to clarify, justify or explain more clearly. The teacher does this by asking the learner to explain more, by asking why the learner thinks s/he is correct, or by asking a specific question that relates to the learner's idea and pushes for something more
Maintain	The teacher maintains the contribution in the public realm for further consideration. She can repeat the idea, ask others for comment, or merely indicate that the learner should continue talking
Confirm	The teacher confirms that s/he has heard the learner correctly. There should be some evidence that the teacher is not sure what s/he has heard from the learner, otherwise it could be press

[1] However, they satisfy some of Nystrand et al'.s other criteria for uptake: they are partly shaped by what precedes them; the student's ideas can change the course of the discussion; and the teacher's next question is contingent on the student's idea, rather than predetermined.

sure that other learners gain access to the learner's thinking. *Elicit* and *press* moves can sometimes seem similar to each other, they are distinguished in similar ways to how Wood (1994) distinguishes focusing from funneling – a *press* move orients towards the learners' thinking, while an *elicit* move orients towards a solution.

Maintain is similar to "social scaffolding" (Nathan and Knuth 2003), and supports the process of learners' articulating their contributions, rather than the mathematics itself (Gamoran Sherin 2002). It is also similar to revoicing (O'Connor and Michaels 1996) and often involves a repetition or rephrasing of the learner's contribution which keeps the idea in the public realm for further consideration. *Insert* is a category that I needed to describe instances when the teacher gave information to learners as a follow up to what they had said. This category is motivated by a similar rationale to that of Lobato et al. (2005), that teachers cannot avoid "telling" and this should be recognised as an appropriate part of their repertoire. I have not described this move as elaborately as Lobato et al. because it did not occur in as many different ways in my data. However, there were certainly instances when teachers did *insert* their own mathematical ideas into the discourse, and the *insert* category enables me to investigate the consequences of this move and its relationship to the other moves.

All five teacher moves function to maintain learner contributions, but they do so in different ways. The main difference between *maintain* and *confirm* and the other three codes is that *confirm* and *maintain* are more neutral, confirming the accuracy of what the teacher has heard or maintaining the contribution very similarly to how the learner said it. *Press, elicit* and *insert* moves attempt to transform the contribution in some way while maintaining it as the focus of the lesson for the class. *Press* moves stay with the learner's contribution, asking for something more and attempting to support the learner to transform her own contribution. *Elicit* moves support the learner to transform a contribution by contributing something new. *Insert* moves are where the teacher transforms the contribution by making her own mathematical contribution. The moves can be arranged on a continuum of less to more intervention as follows: *confirm* is where the teacher makes no intervention, she merely tries to establish what the learner said; *maintain* is where the teacher makes very little intervention, rather she repeats the contribution, in order to keep it going, either for later intervention or transformation, or for other learners to do something with the contribution; *press* tries to get the learner to transform her own contribution by elaborating or justifying it; *elicit* tries to get learners to transform a contribution by contributing something new; and *insert* is where the teacher transforms the contribution by making her own mathematical contribution.

Table 9.2 Subcategories of "follow up"

Teacher	Insert	Elicit	Press	Maintain	Confirm	Follow up
Mr. Daniels	24% (41)	5% (9)	20% (35)	42% (73)	9% (16)	61% (174)
Mr. Nkomo	18% (44)	10% (24)	13% (30)	50% (119)	9% (21)	70% (238)
Mr. Mogale	19% (49)	26% (68)	15% (40)	37% (97)	4% (10)	69% (264)
Mr. Peters	24% (103)	23% (99)	20% (85)	30% (128)	4% (17)	68% (432)
Ms. King	31% (65)	21% (43)	7% (14)	39% (82)	2% (5)	52% (209)

Distributions of Teacher Moves

A first point to note here, is that the category *follow up* accounted for the majority of all five teachers' moves: 61% for Mr. Daniels, 70% for Mr. Nkomo, 69% for Mr. Mogale, 68% for Mr. Peters and 52% for Ms. King. Very few of the alternative codes[2] to *follow up* were evident more than 10% of the time for any of the teachers. It is therefore most useful to focus on the subcategories of *follow up* which are given in Table 9.2, rather than the alternatives to it.

Table 9.2 shows interesting differences among the teachers. The first point of note is that half of Mr. Nkomo's *follow up* moves are *maintain*, which means that he repeats the contribution, asks others for comment or asks the learner to continue. The other four teachers also do substantial *maintaining*, however, they also do more of *elicit*, *insert* and *press*, which suggests that as they follow up learner ideas they work to transform them in some way. Mr. Mogale, Ms. King and Mr. Peters *elicit* more than the other two teachers. Mr. Daniels and Mr. Peters both *press* more than the other three teachers. Also noteworthy is how seldom Mr. Daniels *elicits* and how seldom Ms King *presses*. Finally, all the teachers do a reasonable amount of *inserting*. Subsequent sections will show that the teachers make these moves in response to different learner contributions and they also help to produce different learner contributions with these moves.

The above distributions provide a first level description of pedagogy through teacher moves. The predominance of *maintain* moves in Mr. Nkomo's lessons suggests that he takes a relatively neutral stance to learner contributions. Mr. Nkomo looks different from the other teachers and his pedagogy seems to fit a profile of a less interventionist teacher. How and when he *maintains* contributions in relation to when he *presses*, *elicits* and *inserts* may illuminate the extent of his neutrality. Fewer *follow ups* in Ms. King's lessons might suggest a more 'traditional' teacher. Her predominant *follow up* moves are *maintain* and *insert*, suggesting a possibly interesting mix of some neutrality and some explicit intervention. Mr. Peters and Mr. Daniels both *press* for 20% of the time. Mr. Daniels' *presses* go together with a similar percentage of *inserts* and far fewer *elicits*. Mr. Peters' distribution shows similar percentages of *press*, *elicit* and *insert*.

In the rest of this chapter, I present qualitative analyses of each teacher's moves, showing how s/he responds to the different kinds of learner contributions. The analysis will illuminate similarities and differences among the teachers in this regard, and together with the analysis in Chap. 8, will build towards a trajectory for the emergence of and response to learner contributions.

Mainly *Maintaining*: Mr. Nkomo

Mr. Nkomo maintained learner contributions 50% of the time that he followed up on them, while each of the other moves were evident less than 20% of the time (insert: 18%; elicit: 10% and press: 13%). As discussed in Chap. 8, the predominant learner contributions in Mr. Nkomo's lessons were appropriate errors: 25%, missing information:

---

[2]These were: initiate, inform, direct, affirm and other.

36% complete and correct: 36%. In what follows, I will argue that Mr. Nkomo *maintained* missing information and appropriate errors, sometimes *pressing* as well, trying to get other learners to comment on them. When this did not work, he moved into *elicit* and *insert* moves, funneling learners towards complete and correct answers.

The extracts below come from Activity 2, where the learners had to consider what happens to the graph $y=(x-p)^2$ as p changes. Prior to this, they had been talking about the two specific cases of $y=(x-3)^2$ and $y=(x+4)^2$. In the first extract, Mr. Nkomo initiated a consideration of the more general case. As learners made contributions he followed up on these in different ways.

72	Mr. Nkomo:	So, I wanted to know, what is the effect of p? How does p control the graph in other words? How does p control the graph? Is it doing something to the graph, okay? Eh…let me hear from you	Initiate	
73	Molefi:	p is the turning point		
74	Mr. Nkomo:	p is the turning point	Follow up	Maintain
75	Learners:	Yes		
76	Mr. Nkomo:	p is the turning point, is p the turning point	Follow up	Elicit
77	Learners:	Yes		
78	Mr. Nkomo:	Why do you say p is the turning point?	Follow up	Press
79	Sibu:	When you transfer minus p to the other side its going to become positive		
80	Mr. Nkomo:	It's going to become positive	Follow up	Maintain
81	Learners:	Yes		

Molefi's contribution (line 73) is a missing information contribution. Molefi has seen that p relates to the turning point and possibly sees that it is the x-co-ordinate, but has not spoken about the exact nature of the relationship and the y-co-ordinate. Sibu's contribution (line 79) is an appropriate error. He is trying to explain how the $-p$ in the bracket becomes a positive p as the x-value of the turning point.[3] Mr Nkomo's first response to each of these contributions was a *maintain* move (lines 74 and 80), showing a relatively neutral stance in each case.[4] In the first case Mr. Nkomo continued with an *elicit* (line 76) and *press* (line 78) move, possibly hoping that Molefi would reconsider his response and be more specific about the relationship of p to the turning point. However, Sibu's response to his moves brought in another partial contribution, an appropriate error. Mr. Nkomo did not respond to this beyond his *maintain* move in the extract above.

---

[3] Although some might consider this to be a basic error because it involves incorrect algebraic manipulation, which Grade 11 learners should be able to do, it occurred in all three Grade 11 classrooms, in the context of this part of the task: the relationship of p to the turning point. Given that the underlying concept is about the relationship of the equation to the graph, learners' attempts to understand the relationship were coded as appropriate errors.

[4] This is evident in his tone. The fact that he reacted in the same way to contributions that could count as both correct and incorrect suggests that he was not using the ground rule "repeated questions imply wrong answers" (Edwards and Mercer 1987) although the learners may have been interpreting his moves using this rule.

Immediately after this, Mr. Nkomo initiated a sequence of questions about whether a turning point has one or two co-ordinates, suggesting that he saw the missing information contribution as the one that needed work. It often happened in both Mr. Nkomo's and Mr. Daniels' lessons that missing information contributions and appropriate errors accompanied each other, usually in report backs from previous group work. In these cases, the teachers had to choose which to respond to, and these two teachers usually chose the missing information contributions, as happened in this case. In the extract below, we see how Mr. Nkomo's moves support Kefilwe and Sizwe to complete Molefi's missing information contribution. Sibu's appropriate error was not dealt with at all in the lesson.

94	Mr. Nkomo:	Okay, maybe Kefilwe would like to add from you, on that	Follow up	Maintain
95	Kefilwe:	What he is trying to say is p is the turning point of x. which means p is the axis of symmetry		
96	Mr. Nkomo:	It is the...	Follow up	Maintain
97	Learners:	Axis of symmetry		
98	Mr. Nkomo:	So, p is the axis of symmetry. Okay	Affirm	
99	Mr. Nkomo:	So, p is the axis of symmetry, but p is not the turning point. Agreed	Follow up	Insert
100	Learners:	Yes		
101	Teacher:	Now, how, why but in the meantime its turning at that particular point also, Oh, Sizwe wanted to say something?	Follow up	Elicit
102	Sizwe:	Because Sir, p is the value of the turning point for x		
103	Teacher:	Okay! It is the value of the turning point for...	Follow up	Maintain
104	Learners:	For x		
105	Teacher:	For x okay	Affirm	
106	Teacher:	So, that is very important. What Sifiso is saying, okay. p is the value of the x of the turning point, which is what, which is the axis of symmetry	Follow up	Insert

Here Mr. Nkomo *maintained* a previous contribution by asking Kefilwe to add to it (line 94). She revoiced Molefi's earlier contribution, and completed it (line 95). Mr. Nkomo *maintained* her complete, correct contribution, *affirmed* it and then *inserted* his own position, that while p is the axis of symmetry, it is not the turning point (lines 96, 98 and 99). He then *elicited* the response that he had wanted previously, that p is the x-value of the turning point (line 101), which he *maintained* and *affirmed* (lines 103 and 105), and then summarized with an *insert* move, linking the axis of symmetry to the x value of the turning point (line 106).

The two extracts shown above are similar to and different from each other in important ways. They both show how Mr. Nkomo combined *maintain* moves with the other moves to respond to learner contributions. Mr. Nkomo *maintained* the learners' contributions but then moved quite quickly to other moves to try to develop mathematical ideas. In the first extract, his use of an *elicit* followed by a *press* move allowed for an appropriate error, which distracted from the issue of whether p is the turning point, and which Mr. Nkomo did not *follow up*. In the second extract,

Mr. Nkomo's use of *insert* and *elicit* moves functioned to constrain the discourse and funnel learner contributions to the complete, correct response, that p is the x-value of the turning point and the axis of symmetry.

This pattern was seen often in Mr. Nkomo's teaching. He responded to missing information contributions and appropriate errors with *maintain* and sometimes *press* moves. He worked to complete missing information contributions with *maintain* and *elicit* moves. He *maintained* and *affirmed* complete and correct contributions, and then summarized, with *insert* moves and sometimes with a combination of *elicit* and *insert* moves. Mr. Nkomo's teaching is an interesting combination of revoicing (O'Connor and Michaels 1996) and funneling (Bauersfeld 1980). Mr. Nkomo *maintained* learners' contributions, both correct and incorrect, because he wanted others to hear them and he did seem to be genuinely trying to solicit discussion. However, he resorted quite quickly to *elicit* moves, which served to constrain what learners could contribute. It could be described as a relatively neutral move, together with a more interventionist move combined to create constraining classroom discourse. This pattern could suggest a teacher who was trying to be more reform-oriented and less interventionist, as his interpretation of reform teaching, but who did not yet feel comfortable with using the full range of moves to really hear and engage with his learners' reasoning and ideas. This analysis resonates with Mr. Nkomo's own analysis of his teaching in Chap. 3. The extracts where he struggled to take learners' ideas forward are those where he used predominantly *maintain* moves, trying to maintain his neutrality and was not yet sure how to develop learners' thinking. What this analysis adds is that when he moved away from his neutrality, his discourse became constraining – something that many teachers struggle with.

The Power of *Inserting*: Ms. King

Ms. King used *maintain* and *elicit* moves similar to that of Mr. Nkomo, with the same interesting combination of some neutrality with funneling towards an answer. However, Ms. King also used *insert* moves in ways that none of the other teachers did, explaining important ideas to learners. *Insert* moves are where the teacher's voice is most directly evident and for this reason, it might be seen to be the most "traditional" of the teacher moves; "telling" (Chazan and Ball 1999) rather than asking or pressing. However, when Ms. King explained, she engaged directly with learners' ideas and took them further. I remind the reader that Ms. King's learners had the highest number of complete, correct contributions across the five teachers (59%) and the percentages for the other contributions in her class were: Appropriate errors: 14%, missing information: 10% and beyond task: 9% (see Table 8.2).

In the first extract below, Ms. King read a learner's response to the task: *Can $x^2 + 1$ equal 0?* The response was also written on the board and Clive identified it as his.

| 7 | Ms. King: | This person said it's true if x equals any number, but false if minus one isn't in brackets, he gives us an example, he says, if x is two, two squared plus one is not zero | Follow up | Maintain |
| 8 | Clive: | Its mine | | |

(continued)

9	Ms. King:	But if, he then what is he doing here	Follow up	Maintain
10	Learners:	(*Mutter*)		
11	Learner:	He's squaring the one, but not the negative		
12	Ms. King:	Right	Affirm	
13	Ms. King:	He's squaring the one, but not the negative	Follow up	Maintain
14	Ms. King:	Now, what does, what do you think of that	Follow up	Maintain
15	Gordon:	That's wrong, its wrong		
16	Jimmy:	That doesn't work because when you've got *x* equals minus one, its supposed to be the whole of minus one		

Clive's contribution was an appropriate error, because he struggled to deal with the idea that if $x=-1$ then $x^2=(-1)^2$ rather than $-(1^2)$, hence arguing that if $x=-1$ then the expression could equal zero. Ms. Kings first two *maintain* moves (lines 7 and 9) put Clive's response on the table for discussion. One learner immediately articulated Clive's difficulty (line 11) and again Ms. King *maintained* his contribution, asking the learners what they thought of it (line 14). Jimmy tried to explain why Clive's contribution was incorrect. By using the relatively neutral *maintain* move, Ms. King supported other learners to get to the core of Clive's error and to correct it. Ms. King could usually rely on the strong knowledge of her learners to see each other's errors. However, Clive himself was still not sure and so in the following extract she used *elicit* and *insert* moves and relatively closed questions to funnel Clive to the correct answer.

34	Ms. King:	Yes, you can have a sum minus one times one, but is that *x* squared	Follow up	Insert
35	Clive:	No		
36	Ms. King:	What does *x* squared mean, What does *x* squared mean, Clive	Follow up	Elicit
37	Clive:	That number whatever x is times itself		
38	Ms. King:	That number times itself	Affirm	
39	Ms. King:	So, over here, Clive, if *x* is minus one, tell me what is *x* squared	Follow up	Elicit
40	Clive:	Uh, one		
41	Ms. King:	The long way, the whole thing	Follow up	Elicit
42	Clive:	Minus one times minus one		
43	Ms. King:	Minus one times minus one, What's the answer	Follow up	Elicit

There were a number of similar sequences in Ms. King's lessons. They occurred when she was correcting errors or building from incomplete to complete contributions as well as when she taught new concepts. She worked with *elicit* moves to introduce the concepts and then questioned the learners on them, helping them to relate the ideas to their current knowledge. Similar to Mr. Nkomo, Ms. King's teaching often played out in traditional IRE sequences with relatively constrained questions and responses.

Ms King departed from these constrained sequences when she dealt with more extended complete, correct contributions and beyond task contributions.

In the extract below, Ms. King read a complete, correct contribution to the same task: *Can x² + 1 equal 0?* She invited learners to comment on the contribution with *maintain* and *press* moves, similar to how she invited them to comment on Clive's appropriate error above.

84	Ms. King:	He says, what he's saying is for x squared plus one equals zero, he is saying x squared must be equal to minus one, minus one plus one, but the square of any number can never be negative	Follow up	Maintain
85	Ms. King:	Is that true	Follow up	Press
86	Learners:	Yes		
87	Ms. King:	Does he explain why	Follow up	Press
88	Learner:	No, it can't be		
89	Ms. King:	Lets see, It can't be	Follow up	Maintain
90	Ms. King:	Should he have explained why	Follow up	Press
91	Learners:	Ja		
92	Ms. King:	You think he should have	Follow up	Maintain
93	Learner:	(*inaudible*) It's common knowledge		
94	Learner:	Its common knowledge		
94	Ms. King:	Okay, okay, now this is a very interesting issue	Follow up	Insert
96	Ms. King:	Let's just quickly talk about that, some guys say, he should have, he should have explained, why the square of any number can never be negative	Follow up	Maintain
97	Learner:	But there's no point mam		
98	Ms. King:	He says, he says, it's common knowledge	Follow up	Maintain
99	Learner:	Ja		
100	Ms. King:	That is to do with what we're allowed to assume in a proof, he says, it is so obvious, we can assume it. Other people say no, we need to, just, show it more clearly. It's arguable, alright. So we'll leave it open	Follow up	Insert

In response to Ms. King's *presses* (lines 87 and 89), some learners said that the learner whose contribution was on the board needed to explain why a square number could not be negative (many of them had). The learner concerned argued that this is common knowledge, and did not need to be explained (a beyond task contribution). Ms. King *maintained* the two positions as different sides in a debate (lines 96 and 98), but did not allow the learners to continue with the debate. Rather, with her *insert* move (line 100), she explained the mathematical point that what counts as enough explanation depends on the context and it is acceptable for different people to have different views. It is interesting that she did not choose to allow the learners to have this debate at least for some time, which may have made the point more clearly and she did not work towards a conclusion but rather decided to "leave it open." It may be that she was not sure how to deal with this meta-issue in mathematical discourse.

Ms King was far less hesitant in explaining difficult mathematical concepts that arose in her lessons. In responding to a set of beyond task contributions relating to the concept of i, the square root of -1, Ms. King *inserted* a number of conceptual explanations. Based on an explanation from another teacher in the school, Robert had argued:

> Okay, um. Basically we said that, um, the square root of negative one, squared, equals negative one. And then, you say minus one plus one, then you gonna get nought.

Ms. King *maintained* this contribution, asking other learners to contribute. However, they expressed confusion and Robert struggled to explain the concept. So Ms. King explained:

> Okay. Out of the real numbers system, we talking about real numbers and our assignment referred to real numbers. If you read here: Use a logical argument to convince someone else why the conjecture is either true or false for any real value of x. Now the numbers we study at school are real numbers, numbers on the number line. There are numbers that are off the number line. Okay.

Jonny responded with a question: How could any number not be on the numberline and Ms. King explained further:

> Um, okay, just. If we talk about our Cartesian plane. Now usually we draw graphs or we put points on the Cartesian plane. If you are talking about a point over here, this could be the point three, zero and this could be the point zero, two lets say. Now, if we go off the number line, this here would be the point three, two as a coordinate, okay, as coordinates. Another way of writing this exact thing, is by writing it as three plus two i. So, what in fact I'm saying is, we could think of complex numbers as being points on the Cartesian plane. So, any point you care to name, we would write this as minus five minus two, wouldn't we, you could just as well write it as minus five minus two i, if you wanted to. So, it's one way of representing complex numbers as numbers off the number line.

Ms. King was the only teacher who provided such elaborate explanations of mathematical concepts that went substantially beyond what learners had contributed. Her explanations were clear and well structured, even though unprepared, and responded to learners' contributions. Learners were interested in the issue and asked questions about it and so Ms. King gave spontaneous, clear inputs, which explained important concepts, and enabled learners to ask conceptual questions, to which she responded further. So Ms. King showed a combination of traditional eliciting and explaining, which makes her pedagogy somewhat constraining and interventionist, and at the same time more responsive and conceptual.

In her own analysis of her teaching in Chap. 6, Ms. King shows how her teaching did manage to support all five strands of mathematical proficiency in the classroom, including a substantial amount of conceptual understanding. As argued in Chap. 8, one reason might be her learners' stronger knowledge, so it might be the case that similar strategies would not be as successful with weaker learners. However, based on Ms. King's teaching, I argue that there should be space in classrooms for extended, conceptual explanations of mathematical ideas in response to learner questions. This resonates with the argument of Lobato et al. (2005) that there is a role for telling in the classroom and that it can, when used strategically and appropriately, support learners' conceptual development. The fact that there were so few extended explanations in the other teachers' lessons, suggests that teachers may have been dissuaded from using them, possibly because of reform attempts to discourage them from "telling" (Chazan and Ball 1999).

Strategic Combinations: Mr. Daniels

Mr. Daniels' learners produced the most partial insights (30%), the fewest complete, correct responses (19%), and percentages similar to some of the other teachers for the other contributions: appropriate errors (25%), missing information (11%) and beyond task (11%) (see Table 8.2). In what follows, I show how he both supported and dealt with this range of contributions.

Similar to Mr. Nkomo's lessons, appropriate errors and missing information contributions often emerged together in Mr. Daniels' lessons, usually during report backs, as in the example below. David's report back is a response to the question: *what do you notice as the graph $y = x^2$ is shifted 3 units to the right and 4 units to the left.*

50	David:	If you move it four to the left, then all your x-values will decrease by four, and then, your y-values again will stay the same, your new equation would be y equals x squared minus four		
51	Mr. Daniels:	(*Writes $y = x^2 - 4$ on the board*)	Follow up	Maintain
52	David:	And all, all your, all your coordinates will sit in the second quadrant, therefore, all negative, all your x-values, sorry, will all be negative, and there's again, another pattern, it still increases by one, so it will go negative seven, negative six, negative five, negative four, negative three, negative two, negative one		
53	Mr. Daniels:	David, do you mind just, I just want to see what you mean by that, that there, there's a pattern	Follow up	Press
54	Learner:	Yes		
55	David:	There's a pattern		
56	Mr. Daniels:	Okay, do you mind just drawing the table on that side (*points to board to the right of class*), I just want to see	Follow up	Press

There was a lot to focus on in David's report-back, including missing information and appropriate errors. Mr. Daniels wrote $y = x^2 - 4$ on the board, *maintaining* the appropriate error and suggesting that he had noticed it and wanted to come back to discuss it. He also *pressed* David about the pattern that he was talking about.

Mr. Daniels' two *press* moves are *presses* on missing information as to what the pattern actually looked like. It may be that Mr. Daniels hoped, that as David completed the information on the pattern, he would see that there were some co-ordinates with positive values for x, and correct his appropriate error. However, as David began to draw the table, Mr. Daniels called on another group to present and then began to comment on their contributions. Mr. Daniels did not get back to the pattern, nor to the error of $y = x^2 - 4$ as the equation of the shifted graph.

When appropriate errors and missing information contributions came together, both Mr. Daniels and Mr. Nkomo tended to work first with the missing information contributions. There is evidence that they noticed the appropriate errors and chose to work with the missing information contributions rather than the appropriate errors. They had to make difficult choices about what to deal with because they could not deal with everything that the learners said (see Chap. 10). Both Mr. Daniels and Mr. Nkomo usually decided to move on to the next group before they had dealt with everything that arose in the current report back. This may have been a decision to allow for maximum participation rather than to press on particular responses for an extended period of time (see Chap. 10). It also allowed for groups to build on each other's presentations towards a complete solution, which happened in both classrooms.

I argued in Chap. 8 that the emergence of partial insights in Mr. Daniels and Mr. Mogale's lessons could be related to their learners' stronger knowledge as well as to their pedagogies and that the emergence of beyond task contributions in Mr. Daniels' and Mr. Peters' classes could be linked to their pedagogies. Mr. Daniels dealt with partial insights and beyond task contributions similarly in his lessons, by strategically using a combination of moves to draw them into an ongoing discussion. The analysis below both draws on and supports Mr. Daniels' own analysis in Chap. 4, showing how his pedagogy supported learners' collaborative learning.

In the extract below, Mr. Daniels' class was working on the task: *What happens as the graph* $y = x^2$ *shifts 3 units to the right and 4 units to the left.* Over a number of turns immediately preceding the extract, two learners, Michelle and Lorrayne, had co-produced the question: Why does a negative sign in brackets $y = (x - 3)^2$ correspond to a shift of the graph to the right and a positive turning point; while a positive sign in $y = (x + 4)^2$ corresponds to a shift to the left and a negative turning point. This was a partial insight since they were grappling with the relationship between the sign in the brackets and the sign of the turning point. Mr. Daniels *maintained* the question for the class, signifying its importance as a focal point for discussion.

124	Mr. Daniels:	The question is, they asking, if you look at the equation ok, the graph there, the one on the left hand side, they say that the turning point is negative four and the equation is y equals x plus four all squared, why do I have a negative turning point there	Follow up	Maintain
125	Mr. Daniels:	Am I interpreting your question correctly	Follow up	Confirm
126	Learners:	Yes		

127	Mr. Daniels:	Is that what everybody is struggling with	Follow up	Maintain
128	Learners:	Yes Sir		
129	Mr. Daniels:	What do you think David	Follow up	Maintain
130	David:	Sir they saying, they asking you, why do you have your, why's it a negative turning point when it's a positive		
131	Mr. Daniels:	It's, minus four is the turning point but it's a plus four inside in your equation	Follow up	Maintain
132	David:	Sir, can you say that um it, it could, um (*inaudible*)		
133	Mr. Daniels:	David, don't feel pressured if you can, if, if you	Follow up	Insert
134	David:	I'm thinking		
135	Mr. Daniels:	You thinking, okay	Affirm	
136	Mr. Daniels:	Anybody else will like to try, why do you think, I'm sure that some, I mean you've moved the graph and you've played around with the graph, you must have some idea	Follow up	Maintain
137	Mr. Daniels:	Why, how did you justify it when you were going through the activity	Follow up	Maintain

In the above extract Mr. Daniels made six *maintain* moves. His first and fourth *maintain* moves (lines 124 and 141) re-voiced Michelle and Lorrayne's question. His second *maintain* move (line 127) re-voiced the learners' "yes" that indicated they were also grappling with the same question,[5] while his third, fifth and sixth *maintain* moves (lines 129, 136 and 137) tried to solicit contributions from David and other learners. Mr. Daniels' repeated *maintains* kept the question on the table, maintained a demand for a response and gave the learners some tools with which to respond, without narrowing the mathematical task (Bauersfeld 1980; Stein et al. 1996).

After this learners continued to contribute to the conversation for some time, with a number of appropriate errors, partial insights and missing information contributions. The same missing information contribution came up as in Mr. Nkomo's class, that *p* is the turning point, which Mr. Daniels dealt with somewhat differently. In the following extract, we see Mr. Daniels using *press* and *maintain* moves and then strategically placed *elicit* moves, to take the conversation forward.

169	Mr. Daniels:	Okay, they asking the question, why is the turning point negative	Follow up	Maintain
170	Mr. Daniels:	Okay, Now my question is, what is the turning point there really	Follow up	Press
171	Learner:	It's minus four		
172	Mr. Daniels:	Okay now, what is a point, a point is made up of what	Follow up	Elicit
173	Learner:	*x*- and *y*-coordinates		

(continued)

---

[5] Mr. Daniels had also seen the learners talk about this question in their groups the previous day.

174	Mr. Daniels:	x- and y-coordinates, good	Affirm	
175	Mr. Daniels:	So what is the x-coordinate there	Follow up	Elicit
176	Learner:	It's minus four is your x coordinate		
177	Mr. Daniels:	Good	Affirm	
178	Mr. Daniels:	So what is negative there, the turning point is negative, or is it one of the coordinates that's negative, Okay, let's hear	Follow up	Elicit
179	Winile:	The plus four is not like the x, um, the x, like, the number, you know the x (showing x-axis with hand), it's not the x, it's another number. For that when you do the equation, you get some sense from the answer you get, cause without that p, that minus p, your equation will never make sense		

Having repeated the question (line 169), Mr. Daniels *pressed* on what is the turning point (line 170). His use of "really" suggests that he was looking for something beyond what had already been spoken, which was that the turning point is -4. When a learner repeated "-4," Mr. Daniels used a sequence of *elicit* moves to bring out what he assumed the learners knew, that a point is made up of two coordinates and it is the x-value that is -4. In this case, the *elicit* move functioned to point learners to something they knew but were not using at that point. The learners were talking about the turning point as -4, which is a perfectly appropriate way of speaking about the turning point often used by mathematics teachers and mathematicians. In this case though, Mr. Daniels needed the class to explicitly articulate that a point has two co-ordinates in order that they begin to think about the relationship between them.

Although the above section of the discourse is relatively constrained, it supported Winile to begin to articulate that the turning point is generated from the equation in a more complex way than they had been thinking up until now (line 179). She did not express her ideas very clearly, but her contribution helped the discussion to move to a consideration of the relationship between the x- and y-values in the equation and the turning point, and the role of p in that relationship. So Mr. Daniels' *elicit* moves at this point in the discussion helped to provide an important resource for the learners' thinking.

Mr. Daniels used different combinations of *maintain*, *press* and strategically placed *elicit* moves in order to support learner conversation and development of key ideas. In some ways, the distribution of Mr. Daniels' moves look similar to that of Mr. Nkomo, with slightly fewer *maintain* moves and slightly more of the others. The differences in their pedagogies can be described by the ways in which they used the moves in combination with each other for different, strategically considered purposes. While Mr. Daniels' *elicit* moves[6] might look similar to

---

[6]Another interpretation of Mr. Daniels' *elicit* moves in the above example might be that he was concerned with the formal representation of a point, rather than the substantive relationship between its co-ordinates. I reject this interpretation because of how well timed this move was and how it helped the progression of the discussion.

Mr. Nkomo's, they in fact function very differently, in the contexts of the other moves that the two teachers make.

The above extract also shows how complete correct contributions functioned in Mr. Daniels' lessons. He did not focus on complete correct contributions for their own sake, but rather for what they contributed to learners' ongoing understanding of the task. When a learner produced the correct answer, that the point is made up of x- and y-co-ordinates and that the x-co-ordinate in this case was -4, Mr. Daniels continued to try to focus the learners on the relationship between the two points, which led to Winile's helpful contribution.

So Mr. Daniels used a range of moves strategically to both respond to and support a number of different learner contributions. What the above analysis does not show is that Mr. Daniels managed to generate a conversation among learners for the greater part of one of his lessons. This conversation proved very challenging for both Mr. Daniels and his learners as they struggled to keep track of the key ideas (see Chap. 10 and Brodie, in press). At the same time, it was very generative of partial insights and learners' grappling with important mathematical ideas. Mr. Daniels' learners had relatively strong mathematical knowledge. However, their profile in relation to partial insights, beyond task contributions and complete, correct contributions was very different from the even stronger learners in Ms. King's and Mr. Mogale's classes. Mr. Daniels' pedagogy of using a strategic range of moves to generate conversation was key in supporting partial insights and beyond task contributions.

Supporting Learner Moves: Mr. Mogale

A significant feature of Mr. Mogale's teaching was how he supported learner-learner interaction. These interactions worked differently in his classroom from Mr. Daniels' classroom, they were less lively and more constrained, both mathematically and interactionally. A second difference from Mr. Daniels and similarity to Ms. King and Mr. Nkomo is that Mr. Mogale spent more time in traditional IRE sequences, sometimes funneling learners towards answers. So his pedagogy is an interesting mix of generative discussions among learners, which he supports and constrained teacher-learner interaction to achieve complete correct responses.

The extract below[7] shows how Mr. Mogale used report-backs from group discussion to generate whole class interaction among learners. Mpolokeng was reporting back on the task: *what do you notice as the graph $y = x^2$ is shifted 3 units to the right and 4 units to the left.*

---

[7]The learners spoke in Setswana, this is a translation of the original into English.

182	Mpolokeng:	Ok, the observation is that the values of, of the x values are all negative. They are negative and then they don't cut the y-axis, and our turning point is negative. (*indicates to Pulane to talk*)		
183	Pulane:	What do you mean when you say they don't cut the y-axis?		
184	Learners:	(*mutter*)		
185	Mr. Mogale:	Did you get the question?	Follow up	Confirm
186	Mr. Mogale:	Question again	Direct	
187	Mpolokeng:	I am ok		
188	Pulane:	She says she is ok		
189	Mr. Mogale:	What do you mean when you say they don't cut the y-axis? I am interested in that as well	Follow up	Press
190	Mpolokeng:	Ok isn't it when we take, this graph, they move four units to the left, ne? We moved it, haven't we? Yes, this thing, that graph, that we have moved, it doesn't, it doesn't cut at the y-axis		
191	Mr. Mogale:	Yes, Oratile. And then you Letsapa, Oratile first	Direct	
192	Oratile:	Isn't that the y-axis has arrows at the top of to show that this thing extends. If you extend it, and the graph as well, do you mean that they won't meet somewhere?		
193	Mpolokeng:	They will meet		
194	Oratile:	Will they meet?		
194	Mpolokeng:	Yes		
195	Oratile:	So, when you say it doesn't cut		
196	Mpolokeng:	Yes, but here we did not extend them		

In this extract, Mr. Mogale made only four moves in fifteen turns: a *confirm* move to ask whether Mpolokeng had understood Pulane's question (line 185), a *press* move which indicated his interest in Mpolokeng's statement that the graph didn't cut the y axis, thereby *pressing* her to clarify (line 189) and two *direct* moves, indicating to Pulane to repeat the question and to indicate who should speak (lines 186 and 191).

Mr. Mogale's moves supported the two learners Pulane and Oratile to *press* Mpolokeng to clarify her claim that the shifted graph did not cut the y-axis, an appropriate error. An analysis of learner moves in Mr. Mogale's lessons suggests that they produce similar moves to their teacher, *maintaining* contributions, *pressing* for explanations and *eliciting* new ideas (see Chap. 5 and Brodie 2007a). As Mpolokeng was *pressed* by her peers she made the claim, repeated later by her and another member of her group, that in fact the graph does cut the y-axis if it is extended, but her group did not extend it. This could seem to be defensive, but her insistence later suggests that she was grappling with the idea that the graph did extend to cut the y-axis and so made a contradictory claim, also an *appropriate error*. So discussion among peers raised further *appropriate errors*. This also happened in Mr. Daniels' class, where an extended conversation about the relationship between the sign in the bracket $y = (x+4)^2$ and the sign of the turning point -4, led to a number of additional *appropriate errors*.

In other similar examples, learners in Mr. Mogale's class focused on missing information. Given that Mr. Mogale's learners had strong knowledge, they were able to complete the missing information contributions and correct the appropriate errors that their peers produced. Mr. Mogale's main approach was to encourage learners to challenge each other and resolve the issues together. They often did this, sometimes completing and correcting each other's contributions and sometimes generating partial insights. Mr. Mogale's response to partial insights was similar to his responses to appropriate errors and missing information contributions, he encouraged learners to talk with each other about them. In some instances when learners did not notice the appropriate errors, missing information or partial insights in their peers' contributions, Mr. Mogale himself challenged a learner, thus supporting a discussion around their contribution. His approach was always to *press* and *maintain* until learners came to a resolution.

When learners could not resolve an issue, Mr. Mogale moved into *insert* and *elicit* moves, generating funneling sequences (Bauersfeld 1988), towards a correct answer. In the extract below, we see such a sequence. The class was working on the question: *what changes and what stays the same when the graph of $y = x^2$ shifts 4 units to the left and becomes $y = (x + 4)^2$*. Reagile claimed that the axes of symmetry of the two graphs are different and Mr. Mogale asked him why they are different. The learner answered that it is because the equations are different and the following exchange ensued:

26	Reagile:	The equations are different		
27	Mr. Mogale:	The equations are different	Follow up	Maintain
28	Reagile:	(nods head)		
29	Mr. Mogale:	The equations are different	Follow up	Maintain
30	Mr. Mogale:	As long as you can have a difference in the equations, then they differ	Follow up	Elicit
31	Reagile:	I think so		
32	Mr. Mogale:	(Writes on the board). Okay let's say we have something like y equals, one was $(x+4)^2$, what if we have another one *ya* y equals $-(x+4)^2$. Are these equations the same or different?	Follow up	Elicit
33	Learners:	They are the same		
34	Mr. Mogale:	They are the same	Follow up	Elicit
35	Reagile:	Ee [yes] the value of a, whether it's negative or positive determines the shape of the graph (indicating with his hand). (someone comes into classroom to talk to the teacher)		
36	Mr. Mogale:	The value of	Follow up	Maintain
37	Reagile:	a, determines, okay, the value of a determines the shape of the graph, so (inaudible, indicating with his hand)		
38	Mr. Mogale:	So the value of a determines the shape of the graph	Follow up	Maintain
39	Reagile:	Yes		
40	Mr. Mogale:	So, but are you saying the equations are the same?	Follow up	Elicit

(continued)

41	Reagile:	(thinks, looks doubtful) Yes		
42	Mr. Mogale:	They are the same, if you say they are the same, you simply mean when we substitute our values of x here and here, if we say x is two here and here (points to the two equations on the board) and you simplify, you will come to the same expression, when we simplify. Is that what you are saying?	Follow up	Insert
43	Reagile	(inaudible)		
44	Mr. Mogale	Okay, lets (inaudible) two and negative two, are they the same?	Follow up	Elicit
45	Reagile	No		
46	Mr. Mogale	Why no?	Follow up	Elicit
47	Reagile	The other one is negative (inaudible)		

In his first two turns (lines 27 and 29) Mr. Mogale *maintained* Reagile's claim that the axes of symmetry are different because the equations are different. However, it soon became clear that he wanted to challenge this claim, which he did with *elicit* moves in lines 30 and 32. From the exchange it is evident that the teacher wanted the learners to think beyond the particular case that they were dealing with [$y=x^2$ and $y=(x+4)^2$], where the equations are different, and to think about whether different equations always produce different axes of symmetry. The teacher's question about the two equations $y=(x+4)^2$ and $y=-(x+4)^2$ is a case of the teacher raising the task demands, rather than lowering them (Stein et al. 2000). However, the teacher was so intent on getting the learners to see that different equations do not necessarily generate different axes of symmetry that he ignored Reagile's thinking and subsequently narrowed his own questions in an attempt to funnel Reagile to the correct answer.

Reagile was in fact arguing that the two equations are the same when you consider the axis of symmetry. The only difference in the two equations is the a-value, which determines the shape of the graph rather than the axis of symmetry. Reagile made this argument even in the face of the teacher's disagreement, suggesting that he was convinced of his position. Although the teacher listened to Reagile and *maintained* his contribution in line 38, he did not see it as a contribution to the more general question about differences in the axes of symmetry and so he ignored the gist of Reagile's argument. As the exchange progressed, the teacher narrowed his questions in a number of ways. By focusing attention on the features of the two equations $y=(x+4)^2$ and $y=-(x+4)^2$, he removed attention from their relationship to the graphs, a move that Reagile resisted by continuing to focus on the relationships between the equation and the graph. The teacher then narrowed the question even further to whether 2 and -2 are the same or different, a question that is obvious to any Grade 11 learner, and that is so simple that it completely lost its relevance and context in relation to the original task. In the exchange Reagile was reduced to answering a simple question, rather than having his genuine mathematical thinking taken seriously.

There were a number of similar sequences in Mr. Mogale's lessons suggesting similarities between him, Ms. King and Mr. Nkomo in working towards complete, correct contributions. However, the key difference is that at other times Mr. Mogale enabled strong interaction among learners and supported them to use the same moves

that he made to challenge and support each others' thinking. In this way, his pedagogy was similar to Mr. Daniels' in generating learner discussion and insight. This combination of open and constraining discourse functioned to both support learner–learner interaction and to keep the key mathematical ideas in focus. In Mr. Mogale's own analysis of his teaching in Chap. 5, he focuses on his moves and shows that his learners internalized these and used them in the discussion. Mr Mogale had worked hard with these learners to develop norms of participation in classroom discussion, and the analyses here and in his chapter attest to his success. He also noted in his own analysis, that he sometimes shifted from listening to learners, to listening for particular answers, so he was aware of when his pedagogy shifted. What these analyses show is that even after working for some time with learners to develop robust forms of interaction, it is likely that teachers will slip into familiar patterns and undermine some of their own attempts. This is to be expected as teachers take on the difficult task of shifting their practices (see also Chap. 11).

Entertaining Errors: Mr. Peters

Mr. Peters' learners produced basic and appropriate errors in 40% of their contributions. Therefore, much of his work entailed recognizing and working with these errors. In Chap. 8, I argued that Mr. Peters' response to both basic and appropriate errors was to ask learners to explain how they got them and to keep them on the agenda for discussion for some time. In the extract below, Ahmed made two incorrect claims: that $x=1$ because there's a 1 in front of the x (line 31)[8] and that $x^2+1=2x^2$ (line 41). It is not clear whether Ahmed sees these claims as related.

31	Ahmed:	Like you always say sir, in front of the x you see a one		
32	Mr. Peters:	In front of the x you see a (pause)	Follow up	Maintain
33	Learners:	One		
34	Mr. Peters:	A one, here	Follow up	Maintain
35	Ahmed:	So um, you plussing one still, sir		
36	Mr. Peters:	But this	Other	
37	Ahmed:	You can get an answer		
38	Mr. Peters:	So, what is the answer	Follow up	Press
39	Ahmed:	Um, if you have one x squared plus one, sir		
40	Mr. Peters:	What's one x squared plus one	Follow up	Press
41	Ahmed:	If I add it, if I add it sir, one x squared, would give me, plus one would give me two x squared		
42	Mr. Peters:	Would give you two x squared right (pause), so x squared plus one, because of the one there, Ahmed says, that's gonna be equal to two x squared (writes on board $1x^2 + 1 = 2x^2$). Has anyone got something to say about this	Follow up	Maintain

[8] This was in response to another learner's previous contribution that $x=1$.

In the above extract, Mr. Peters did not move to correct Ahmed's errors. Rather he *maintained* them by getting Ahmed to repeat the first error in lines 32 and 34 and by repeating the second error himself in line 42. He also *pressed* Ahmed for an "answer" to $x^2 + 1$ in lines 38 and 40, supporting the second error. After this extract, other learners indicated that they agreed with Ahmed, while some argued that you "cannot add like terms." Mr. Peters built on these contributions in order to correct the error, with predominantly *elicit* and *insert* moves, in the extract below.

66	Teacher:	What is $a + b$ equal to? These are unlike terms. What is $a + b$ equal to?	Initiate	
67	Class:	$a + b$ (*chorus*)		
68	Teacher:	Why can't I say one plus one? There's a one and there's a one in front there. Can't I say two ab?	Follow up	Insert
69	Class:	No, sir (*chorus*)		
70	Teacher:	Why not?	Follow up	Press
71	Class:	Unlike terms (*chorus*)		
72	Teacher:	Put up your hands lets see. Victor, I'm going to write what you gonna tell me	Direct	
73	Victor:	*Silence*		
74	Teacher:	Grace	Direct	
75	Grace:	Because they unlike terms…		
76	Teacher:	So can I do that?	Follow up	Elicit
77	Class:	No, sir (*chorus*)		
78	Teacher:	So is one x squared plus one equal to two x squared?	Follow up	Elicit
79	Class:	No, sir (*chorus*)		

Here we see Mr. Peters teaching the rule that you can't add unlike terms. He does not explain the reason behind the rule, rather he gives an example ($a + b$) and then relates it to the example on the board (line 78). This extract has a very different tone and pace from the previous one. Here Mr. Peters clearly wants specific answers from the learners, hence the *insert* and *elicit* moves, rather than to hear what the learners are thinking as in the previous extract. Learners' consistently *chorused* responses (lines 67, 69, 71, 77 and 79) show a co-produced narrative of the "right story." However, it is unlikely that Ahmed, or others, have achieved a deeper understanding of why you can't add unlike terms.

Mr. Peters followed the same pattern with all the other basic errors: He gave them attention by using *maintain* moves, tried to understand learners' reasoning using *press* moves, and then worked to correct them using *elicit* and *insert* moves. His moves to correct the errors were always relatively procedural, even though it is clear from his interviews that Mr. Peters has a conceptual understanding of mathematics. Many of the *basic errors* related closely to Mr. Peters' lesson agenda of dealing with the *appropriate errors* and *partial insights* that learners had generated. The way he worked with *basic errors* enabled him to both deal with them, and come back to the main mathematical issue under discussion, so that the *basic errors* did not detract too much from his mathematical agenda for the lesson.

Mr. Peters worked to *maintain* appropriate errors for discussion and to *press* learners to clarify their thinking in a similar manner. In the extract below, he asks Grace to explain her contribution that $x^2 + 1$ cannot equal zero because it cannot be simplified.

3	Mr. Peters	Grace, do you want to say something about that, what were you thinking, what were the reasons that you (*inaudible*)	Follow up	Press
4	Grace	Sir, because the x squared plus one ne sir, you can never get the zero because it can't be because they unlike terms. You can only get, the answers only gonna be x squared plus one, that's the only thing that we saw because there's no other answer or anything else		
5	Mr. Peters	How do you relate this to the answer not being zero, because you say there it's true, the answer won't be zero, because x squared plus one is equal to x squared plus one. You say they're unlike terms. Why can't the answer never be zero, using that explanation you are giving us	Follow up	Press
6	Grace	(*Sighs and pinches Rethabile*)		
7	Mr. Peters	Rethabile, do you wanna help her	Follow up	Press
8	Rethabile	Yes, sir		
9	Mr. Peters	Come, let's talk about it	Direct	
10	Rethabile	Sir, what we wrote here, I was going to say that the x squared is an unknown value and the one is a real number, sir, so making it an unknown number and a real number and both unlike terms, they cannot be, you cannot get a zero, sir, you can only get x squared plus one		
11	Mr. Peters	It can only end up x squared plus one	Follow up	Maintain
12	Rethabile	Yes, sir, there's nothing else that we can get, sir, but the zero, sir		
13	Mr. Peters	So you can't get a value, you can't get a value	Follow up	Maintain
14	Rethabile	That's how far we got sir		
15	Mr. Peters	Come, Lebo, lets listen so you can contribute	Direct	
16	Mr. Peters	So it will only give you squared plus one, it won't give you another value, zero. Will it give us the value of one, will it give us the value of two, x squared plus one	Follow up	Press

In the transcript we see Mr. Peters *pressing* for explanations (lines 3, 5, 9), *maintaining* the girls' claim (lines 11 and 13) and trying to *press* them more specifically to think about the expression (lines 5 and 15). His *press* moves ranged from being very general (line 3) to more specific (line 16). However, even this final question which could have supported learners to think about $x^2 + 1$ as taking several values depending on the value of x, did not help. In fact, it led Rethabile directly into one of the *basic errors* discussed above (x is equal to 1).

In all the cases of *appropriate errors* Mr. Peters' response was to work on them in two ways. First, he kept them in the public arena for discussion, held up different ideas for discussion for some time and tried to get learners to justify and clarify their thinking. He did not always move to teaching the correct answer as he did with *basic errors* and when he did so, he worked less procedurally. Second, he planned new tasks, which he hoped would help learners with their errors (see Chap. 8 for more detail). However, each time Mr. Peters began a discussion on an *appropriate error* a host of *basic errors* came up. Mr. Peters dealt with these *basic errors* relatively quickly as discussed previously and then came back to the discussion of the *appropriate errors*. Thus, Mr. Peters both supported and noticed learner errors and focused on discussing them.

Mr. Peters' responses to the other kinds of contributions were similar. In the case of missing information contributions he maintained and pressed, and then moved to complete them with *elicit* and *insert* moves, similarly to how he worked with basic errors. With partial insights, he *maintained* and *pressed*, keeping them on the agenda for consideration, similarly to appropriate errors. There were very few instances of extended complete and correct answers in his lessons and he *pressed* for explanation and justification on those that there were, similar to the other teachers. In the case of the beyond task contributions, he developed a conversation using all the moves (Brodie 2007b). In the case of appropriate errors and partial insights, Mr. Peters did not move towards complete, correct responses immediately. Rather, his response in the first two cases was to plan more tasks that would help learners to get to the reasons for their errors and insights (see Chap. 8).

Mr. Peters' classroom looked different from all of the other teachers in that there were so many errors. Mr. Peters found these somewhat disturbing (see Chap. 7), as would many other teachers. However, if we take the misconceptions literature seriously, particularly that errors are an important part of learning, then errors need not be seen as problematic. Rather, this analysis suggests that Mr. Peters was succeeding somewhat in engaging the learners' thinking, in that he supported them to express their incorrect ideas and engaged with them as an important step towards correct understandings.

Overview: Teacher Responses to Learner Contributions

The analysis up till now has highlighted similarities and differences among the teachers' responses to learner contributions. Tables 9.3 and 9.4 provide summaries of this analysis.

Here we see strong similarities between Mr. Nkomo and Mr. Daniels, which, as discussed in Chap. 8, are task-related. Their tasks produced similar missing information and appropriate error responses, particularly in report backs from group work. Because there was usually so much to deal with at any one time, the teachers had to make decisions as to what to take up and what to ignore, and both of them tended to focus on the missing information contributions. They *maintained* and *pressed* these,

Table 9.3 Teacher moves and learner contributions (part 1)

	Basic errors	**Appropriate errors**	**Missing information**
Mr. Nkomo	None	When dealt with, used *maintain* and *press* to discuss. Missing information took precedence	*Maintained* and *pressed*. Used *elicit* and *insert* to complete
Mr. Daniels	None	When dealt with, used *maintain* and *press* to discuss. Missing information took precedence	*Maintained* and *pressed*. Different contributions came together to complete
Mr. Mogale	None	Supported learners to question using *press* and *maintain*. Sometimes corrected using *elicit* and *insert*	Supported learners to question using *press* and *maintain*. Sometimes corrected using *elicit* and *insert*
Ms. King	None	Used *maintain to* discuss. Corrected using *elicit* and *insert*	*Maintained* and *pressed*. Used *elicit* and *insert* to complete
Mr. Peters	Used *maintain* and *press* to understand. Used *elicit* and *insert* to correct	Worked to understand using *maintain* and *press*. Planned new tasks to deal with	*Maintained* and *pressed*. Used *elicit* and *insert* to complete

Table 9.4 Teacher moves and learner contributions (part 2)

	Partial insights	**Complete correct**		**Beyond task**
		(short)	**(long)**	
Mr. Nkomo	Only one (not discussed here)	Summarised using *insert* and *elicit*		None
Mr. Daniels	Promoted discussion using all moves	Strategic use of all moves to help discussion and to take contribution further		Developed discussion using *all moves*
Mr. Mogale	Supported learners to question using *press* and *maintain*	Built towards using *elicit* and *insert*	Supported learners to justify, using *press*, *maintain* and *elicit*	None
Ms. King	Only two (not discussed here)	Built towards using *initiate* and *elicit*	Took further using *insert* and *elicit*	*Maintained* and *pressed*, then explained new math using *elicit* and *insert*
Mr. Peters	Used *maintain* and *press* to engage. Worked into planning	Used *elicit* and *insert* to correct basic errors and to complete missing info contributions	*Pressed* for justification and explanation	Developed discussion using *all moves*

and called on other groups to try to complete them. Mr. Mogale, who used the same tasks as Mr. Nkomo and Mr. Daniels, and whose learners made the same appropriate errors and similar missing information contributions, responded somewhat differently. His pedagogy consistently supported learners to challenge each other, and because of his learners' stronger knowledge, they were able for the most part to recognize and challenge their peers' errors and missing information contributions. Mr. Mogale's learners had internalised many of his moves and used them in the classroom conversation, thus supporting his work with them. Even though Mr. Nkomo and Mr. Daniels encouraged learners to comment on each other's solutions, their challenges often did not generate useful consideration of the mathematical issues.

We also see interesting similarities and differences between Ms. King and Mr. Peters. They both *maintained* and worked with the appropriate errors, although in different ways. Ms. King moved quite quickly to correct these, and she was helped to do this by the many learners in her class who could recognise and correct their peers' errors. Mr. Peters *maintained* appropriate errors for longer periods of time, using them to try to teach the important concepts and working them into his planning. His practice shows that he viewed these errors as important to work with in extended ways. It is interesting that Mr. Peters and Ms. King worked similarly in relation to *basic errors* and *appropriate errors* respectively. They *maintained* these, tried to understand them and then moved to correct them. These may be appropriate strategies in relation to their learners' knowledge. Mr. Peters and Ms. King worked in similar ways in relation to *missing information* contributions; they both *maintained* them and then worked to complete them.

I argued in Chap. 8 that *partial insights* were strongly related to the teachers' pedagogies. They occurred in the classrooms where the teachers managed to generate discussion among learners for at least some of the time. *Partial insights* reflect the fact that learners are grappling with important ideas and this study suggests that to get them to do this, at least publicly, there needs to be some discussion, even discussion which is not yet totally successful. Both Mr. Peters and Mr. Daniels used *partial insights* to spur further thinking and discussion among learners. Mr. Peters missed a number of *partial insights*, for the same reasons that Mr. Nkomo and Mr. Daniels missed some appropriate errors: Once the lesson is opened to learner ideas, many ideas come up; sometimes too many to deal with at once. At the same time, those *partial insights* that Mr. Peters thought were important, came to play a prominent role in his lessons, forming much of the lesson agenda and influencing his further planning, and in so doing, brought learners' thinking into subsequent tasks and lessons.

In relation to *complete, correct* responses, Table 9.4 shows that all the teachers took these further in some way; they did not merely accept them as correct. Mr. Nkomo summarized them for the benefit of the class, Mr. Daniels, Mr. Mogale and Mr. Peters pressed for more elaboration and justification and Ms. King used them to teach further mathematics. All of the teachers worked towards *complete correct* contributions using similar moves but for slightly different purposes. Mr. Nkomo used sequences of short questions in order to summarize learner contributions and to teach the general concepts that learners did not develop through his other moves. Mr. Daniels used these to point to important mathematical points that might help learners' thinking and take the discussion further. Ms. King, Mr. Peters and Mr. Mogale all worked to correct errors and complete missing information contributions. All of the teachers except Mr. Daniels worked towards short complete correct responses using the constraining questioning and funneling of the traditional IRE. This is an important part of teachers' work and cannot be ignored. At the same time, all of the teachers knew that correct responses do not necessarily signal the end of a discussion, they can be taken further and all of the teachers did take them further, particularly the extended contributions that came out of report backs.

Contributions that went beyond the task came up in three of the five classrooms and were dealt with in different ways. Ms. King used these to explain new mathematical ideas. Mr. Daniels and Mr. Peters both used these to try to generate some discussion among learners. These discussions allowed a range of ideas to be exchanged, and provided for genuine mathematical development on the part of learners in Mr. Daniels' class. They also provided real challenges for these teachers, some of which will be discussed in Chap. 10.

Trajectories for Working with Learners' Contributions

The analyses in Chaps. 8 and 9 come together to suggest possible trajectories for the emergence of and responses to learner contributions. This is discussed below and reflected graphically in Fig. 9.1.

All of the teachers worked to obtain *complete, correct* contributions. It is difficult to imagine teaching without such an orientation to *complete correct* responses, except in the most *laissez faire* child-centered teaching approaches, which are highly unlikely to exist in secondary mathematics classrooms (Chung and Walsh 2000; Cuban 1993), and would be highly inappropriate. An important question for teachers is how many of these do they need to see in their classrooms in order to feel that they are making progress, or could some of the other contributions count as progress towards mathematical ideas, as they did in these five classrooms. All of the teachers took *complete, correct* responses further, requiring more of the learners who produced these. In addition, the teachers worked with *complete, correct*

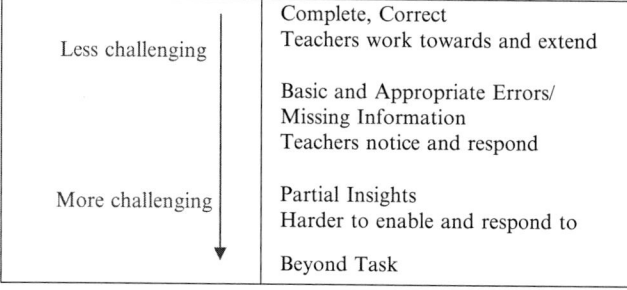

Fig. 9.1 Emergence of and response to learner contributions

responses for different purposes. This suggests that one way of beginning to work towards reform pedagogies is for teachers to think about ways in which *complete, correct* contributions can be taken further, becoming the beginning points rather than the end-points of conversations. Given that teachers will work towards *complete, correct* contributions and will most likely have criteria for recognizing these, this might be a first step in a possible trajectory towards more reform-oriented teaching, a first step which builds on teachers' strengths and helps them to develop new ways of teaching.

My analysis also shows that while these teachers worked towards *complete correct* responses, they also worked extensively with other kinds of contributions. They worked to maintain *basic* and *appropriate errors* and *missing information* contributions so that they could understand them, and only then worked to correct and complete them. They also tried to get other learners to help discuss and work with these. So they brought learner ideas into the public space of the classroom accepting them as valid ideas, while working to develop and transform them. They saw *basic* and *appropriate errors* and *missing information* as reflecting both presences and absences in learner thinking and worked with these in a variety of ways, constrained particularly by the tasks and their learners' knowledge. My analysis suggests an important challenge for teachers working with *basic errors, appropriate errors* and *missing information* contributions. The reform literature in general, and the misconceptions literature in particular, suggest that teachers need to understand errors or partial contributions and raise them for discussion in class. The final step, as to how they are actually corrected is never explicitly discussed, almost as if it will occur spontaneously. My analysis shows that these next steps are likely to play out in different ways in different classrooms – in patterned ways that are related to tasks, learner knowledge and pedagogy.

This analysis has also suggested different ways to think about the three subcategories of partial contributions: *missing information, appropriate errors* and *partial insights*. Based on the analysis in this chapter, *partial insights* can be seen differently from *appropriate errors* and *missing information* contributions. They occurred in only three of the five classrooms and where they occurred, the teachers worked with them to generate and maintain discussion. In the case of *appropriate errors* and *missing information* there was emphasis on both what the learners were thinking and at the same time, pressure to complete them and take them towards correct contributions. Thus, *partial insights*, while harder to generate, may be more worthwhile in terms of developing learner thinking in that they reflect connections and integration across ideas. My analysis suggests that partial contributions be grouped into two categories, one being *appropriate errors* and *missing information* contributions, and the other *partial insights*. Depending on a teacher, teacher-educator's or researcher's priorities, these can be worked with in different ways. For example, some might seek to develop ways of understanding and making public, *appropriate errors* and *missing information* contributions before *partial insights*, given that these may be easier to work with.[9] Others might want to focus on the more difficult category of *partial insights,* given their importance in developing mathematical thinking.

Finally, I have argued that three of the five teachers saw *beyond task* contributions as important presences in their classrooms and supported these when they came up. I have argued that the fact that they came up in the three classrooms can be accounted for by a combination of the tasks, learner knowledge and pedagogy, which also shows why they did not come up in two classrooms.

So, the analysis shows why *complete correct* contributions, *basic* and *appropriate errors* and *missing information* contributions, emerged more easily in the five classrooms and how the teachers worked more easily with these, and why *partial insights* and *beyond task contributions* are more difficult to achieve and work with. This suggests possible trajectories for teachers learning to shift their teaching in the direction of reform-oriented pedagogies. I have suggested that shifting ways of working with complete, correct contributions might be a place to start, given that this is an area of strength for teachers. Working with basic errors where they occur and/or partial contributions would be next, beginning either with *partial insights* or *appropriate errors* and *missing information* contributions, depending on particular priorities. *Beyond task* contributions could come before partial responses or after them in the trajectory, again depending on priorities in particular situations. This trajectory is graphically depicted in Fig. 9.1.

As I analysed the teachers' responses to the learner contributions, I pointed to a number of challenges that the teachers experienced in dealing with their learners' thinking and reasoning. These include choosing which responses to deal with, how to complete and correct contributions, how long to stay with particular learners' ideas and how to conclude conversations. These will be discussed in more detail in the next two chapters.

---

[9] Based on the analysis, I can also argue that in the context of certain tasks (i.e., the inductive, compare and contrast type task), missing information contributions seem to be easier to deal with than appropriate errors.

Chapter 10
Dilemmas of Teaching Mathematical Reasoning

In the previous two chapters I developed codes for learner contributions and teacher moves, which enabled a micro-ethnographic analysis across the five classrooms. The analysis illuminated patterns in how the teachers responded to the learner contributions and suggested possible trajectories for the emergence of learner contributions in classrooms and teachers' responses to these. I now turn to a more in-depth analysis of two issues that arose in some of the classrooms: dilemmas of teaching mathematical reasoning, which I discuss in this chapter and resistance to changing patterns of teaching, which I discuss in the in the next chapter.

In this chapter, I look in more detail at two of the classrooms, Mr. Peters' and Mr. Daniels' classrooms. I identify two dilemmas that these two teachers experienced as part of their teaching mathematical reasoning to their learners. These dilemmas relate to other dilemmas in the literature, in both traditional and reform contexts. They were also experienced in some form by the other teachers in this study. Key to these dilemmas in these classrooms is the *press* move. I will show how the *press* move is implicated in two dilemmas of teaching mathematical reasoning as experienced by Mr. Peters and Mr. Daniels.

Teaching Dilemmas

Lampert (1985) defines teaching dilemmas as situations where teachers are confronted by equally undesirable alternatives because of the varied and contradictory aims of the work of teaching. Using examples from her own teaching, she shows how dilemmas do not present clear and obvious choices for teachers, neither in the moment of teaching, nor with considerable reflection afterwards. In fact, Lampert's careful articulation of her dilemmas shows exactly how intractable they can be. She argues that teachers cannot be guided by research or teacher education to make better choices among dichotomous alternatives or to resolve dilemmas. Rather,

---

Chapter 10 is reprinted in part from Journal of Curriculum Studies, K. Brodie, Pressing Dilemmas: meaning-making and justification in mathematics teaching. Copyright (2009), with permission from Taylor and Francis. http://www.informaworld.com

dilemmas are an inherent part of teaching and the conflicts that they present are constitutive of the practice of teaching. For Lampert, "the conflicted teacher is her own antagonist, she cannot win by choosing" (p. 182). Rather, the teacher has to accept the inner struggle that dilemmas bring, manage dilemmas in ways which may not resolve them but which enable further action, and through deliberation and practical reasoning manage to "act with integrity while maintaining contradictory concerns" (p. 184). A review of the literature on teaching dilemmas shows that dilemmas fall into two main categories: those that involve teachers managing the tension between learners' current knowledge and the subject matter they are teaching and those that involve teachers managing tensions between individuals and the class. In practice, these are experienced simultaneously, but it is useful, at least at first, to separate them analytically.

Linking Learners with the Subject

Edwards and Mercer (1987) and Jaworski (1994) identify the "teacher's dilemma" as a recurring aspect to reform pedagogy. This dilemma arises in classrooms where the teacher wants both learner's to participate and to teach particular ideas. The dilemma is how to elicit the knowledge from learners that she wants to teach. As long as the teacher genuinely allows learners to express their thinking, she cannot be sure that it will help to build towards what she is trying to teach and if learners divert significantly from her agenda, she will have to work harder to bring them back. If the teacher maintains her focus on covering the content of the curriculum, then she is in danger of losing out on what learners have to say and making connections between their meanings and the new knowledge.

Ball (1993, 1996, 1997) articulates this dilemma as how to respect both the integrity of the students' thinking and the integrity of the mathematics they need to learn. This dilemma manifests particularly when learners make mistakes. For example, when her students argued that 5/5 is bigger than 4/4 because 5/5 has more pieces, Ball had to work out how they were thinking about fractions in order to make this claim and how to work with them to understand the equivalence of 5/5 and 4/4. Research on misconceptions (Confrey 1990; Smith et al. 1993) suggests that misconceptions are important steps en route to correct mathematical knowledge and that teachers can infer the reasoning and sense behind learners' errors in order to help build new knowledge from old. For the teacher, the challenge is how long to allow errors to persist publicly in the classroom, and exactly when and how to step in and challenge them. Taking learners' ideas seriously when they are mistaken requires that the teacher work out when and how to intervene to both value and validate the learners' contributions and to develop appropriate mathematical knowledge.

In learning to teach in a reform-oriented way, Heaton (2000) describes, when she asked for student contributions, how many were not at all helpful and she struggled to work out what to do with them. Her dilemma was how to gain maximum participation from all learners while simultaneously taking the mathematics under discussion forward and developing mathematical ideas. Heaton learned to manage this dilemma by learning to take control of the discussion while still taking students' ideas seriously. In order to do this, she had to learn to understand the mathematical purpose of the task and to recognize the mathematics that related to this purpose in the students' contributions. Only then could she intervene appropriately in ways that enabled her to teach mathematics while drawing on students' contributions.

A related concern is how teachers hear what learners are saying (Ball 1997; Davis 1997; Wallach and Even 2005). Hearing the ideas of others always requires some form of interpretation against one's own perspectives. Since teachers have curricular goals to accomplish, they are likely to hear what learners say in relation to what they are trying to achieve, what Davis (1997) calls evaluative listening; rather than in relation to the thinking that underlies what learners are trying to say, what Davis calls interpretive listening. Understanding learners' meanings in ways that enable teachers to support learning requires that teachers move beyond their own expectations and understandings and even their own mathematical knowledge (Chazan 2000; Heaton 2000), to really understand the meanings that learners make of mathematical ideas, even when they are not correct (see also Schifter 2001).

Working Simultaneously with Individuals and Groups

All teachers experience the tension between needing to develop individual learners' understandings while teaching a whole class. In traditional teaching, this is often experienced as managing different ability levels and personalities among learners. In reform teaching, these challenges take a different form, as teachers endeavour to engage learners with each other's thinking in order to create communities in which learners might further develop their ideas. Developing mathematical communities in classrooms requires that teachers support all learners to express their ideas and participate in the discussion and that they mediate appropriately between different learners' ideas.

Chazan and Ball (1999) describe how they work to create classroom mathematical communities and how these break down in instances of what they call "unproductive agreement" and "unproductive disagreement". Unproductive agreement occurs during discussion when all students agree that an error is correct, and do not challenge it. Unproductive disagreement occurs when students disagree so vehemently with each other that very little genuine thinking or reconsideration of ideas can take place. Chazan and Ball found that they could not rely solely on students to challenge each other appropriately and take the mathematics forward, since group dynamics and relationships among students over-determined the conversations. They argue that the role of the teacher is to find appropriate ways to insert his or her voice to calm overexcited or defensive argumentation, or to stir up challenge where necessary. Osborne (1997) describes the case of a learner, Cory, who is both productive and disruptive. Cory is extremely insightful and, therefore, helpful in class in that he raises ideas that provoke thinking among his peers. At the same time, his behaviour can be so disruptive that he distracts others from thinking. Osborne describes how she walks a fine line between disciplining him and thus preventing him from contributing, and disciplining him so that he and others can contribute.

Two key decision points for teachers are who to call on to contribute at a particular point in the conversation and when to call on particular learners. Most teachers want to give equal opportunities to learners to express their ideas, and yet in all classes, particular learners can be more or less helpful at particular times in taking discussions forward. An experienced reform teacher, Lampert (2001), chooses particular students to contribute at particular times for mathematical reasons "I used my choice of whom to call on to get a particular piece of mathematics up for consideration" (p. 146). In researching the practices of a successful reform teacher, Boaler argues that strategic choices of student contributions at particular times can help shape the mathematical direction of the class. Since all students are likely to have ideas to offer at different times, making strategic choices could also include making equitable participation possible over longer time periods (Boaler and Humphreys 2005). These authors argue that managing the tensions between individuals and the class requires that the mathematical purposes of the lesson be foregrounded. While it is clearly important for all learners to contribute, learner contributions can be organized in ways that take the mathematical conversation forward, and this organization of learner contributions is a key role for the teacher in reform classrooms.

The above review of research across a number of contexts suggests that there are similarities in the dilemmas that teachers experience. At the same time, Lampert (1985) points to the deeply contextualized nature of teaching; dilemmas arise in particular ways in particular classrooms, which are experienced differently by teachers and are managed by teachers in locally conducive ways. In Mr. Peters' and Mr. Daniels' classrooms, the dilemmas discussed above were experienced in relation to how the two teachers used the "press" move to engage with learners' mathematical reasoning. I identify two dilemmas experienced by Mr. Peters and Mr. Daniels: whether to press or not on particular learners' meanings and whether to take up or ignore learner contributions. I show how the different choices that the two teachers made as to how they pressed their learners' contributions had different consequences for how they experienced the dilemmas of teaching mathematical reasoning.

The "Press" Move

As discussed, in Chap. 9, pressing on learner's ideas is a reform-oriented practice, where the teacher asks the learners to elaborate clarify, justify, or explain their ideas more clearly. Teacher presses can range from the general "can you say more" to very specific presses "why did you choose 2?" Other examples are "what do you mean by …" and "can you elaborate". Teacher presses can be distinguished from the more predominant "elicit" move (Edwards and Mercer 1987; Mehan 1979) which is common in traditional classroom discourse, in that teacher presses aim to enable the learner to transform her own contribution and thinking, rather than to produce a different contribution, usually the "correct" answer from the teacher's perspective. Elicit moves often serve to narrow the discourse and "funnel" the learner's contributions towards a particular point (Bauersfeld 1980), whereas press moves try to stay with the learner's thinking.

Pressing shifts the norms of classroom discourse, in that it does not comply with the commonly accepted ground rule identified by Edwards and Mercer (1987) that repeated questions imply wrong answers. A teacher may press on a correct answer, if she wants the learner to elaborate the idea more clearly, either for the teacher, the learner himself or herself or other learners. Pressing also violates a pervasive classroom norm that teachers only ask questions to which they already know the answers. Teachers might press with authentic questions (Nystrand et al. 1997) if they genuinely do not understand how a learner is thinking and want to find out. They can also press when they do know what a learner means and want the learner to re-articulate the idea or provide a justification. These shifting and more complex ground rules make reform classrooms harder to negotiate for both learners and teachers.

Kazemi and Stipek (2001) distinguish between low and high presses in mathematics classrooms. High presses require that when learners explain their thinking they provide a mathematical argument underpinned by conceptual relationships while low presses accept procedural explanations of problem solving strategies. In this chapter, I distinguish between presses for meaning and presses for justification. The first transcript below shows a press for meaning and the second a press for justification (the presses are in bold font). David was reporting on his group's discussion on the task: what happens when the graph of $y=x^2$ is shifted four units to the left. This was the first part of a task that asked students to explore differences in the graph and equation as $y=x^2$ shifted horizontally and vertically.

David:	And all, all your, all your coordinates will sit in the second quadrant, therefore, all negative, all your x-values, sorry, will all be negative, and there's again, another pattern, it still increases by one. So it will go negative seven, negative six, negative five, negative four, negative three, negative two, negative one.
Mr. Daniels:	**David, do you mind just, I just want to see what you mean by that, that there, there's a pattern**
Learner:	Yes
David:	There's a pattern
Mr. Daniels:	**Okay, do you mind just drawing the table on that side (*points to board to the right of class*), I just want to see ...**
Mr. Daniels:	**My question to them is just, why did you do what, why did you do that**
Ntabiseng:	So everything can be clear, sir
Mr. Daniels:	Sorry
Ntabiseng:	So everything can be clear
Mr. Daniels:	What do you mean by everything can be clear
Learner:	When you move
Ntabiseng:	Ja, you can see, how you move, and stuff like that

The presses in the first transcript focus on what the learner means by a pattern. In this case, Mr. Daniels' presses were authentic, he did not understand where David saw a pattern and wanted him to elaborate for himself and the other learners. In the second transcript, the first press is a press for justification. The learner's response to this press did not provide an adequate justification, hence Mr. Daniels' subsequent press, which looks like a press for meaning but functions as a press for further justification. It is likely that in this case Mr. Daniels had an idea as to why the learners had shifted the graph, but he wanted them to clarify this for the rest of the class, and for themselves.

In both of the examples above, and in many other examples across the five classrooms, teacher presses often did not succeed in producing clearer, more articulate ideas or evidence of deeper or transformed thinking on the part of the learner (Brodie 2005) and, therefore, might be considered examples of low presses by Kazemi and Stipek. In this chapter, I try to understand some of the reasons for this finding, which are located in the difficulties associated with the press move and the dilemmas it creates for teachers.

To Press or Not to Press?

As discussed above, a common challenge for teachers in reform classrooms is when and how to intervene to link learners' ideas with important mathematical ideas and to develop them further. The dilemma of whether to press or not arises once the teacher has made a decision to intervene and is faced with decisions about how to do so: whether to continue pressing a particular idea or learner and for how long. The concerns that arise in relation to how long to press relate to the two dilemmas discussed above: working between learners' ideas and the subject and working between individuals and the group. While pressing can help the teacher to understand a particular learner's thinking, it may divert the class from the mathematics under discussion, and it also may reduce the time available for others to contribute.

Across the five teachers, Mr. Peters was more comfortable than the others in pressing longer on learners' thinking, and I use an example from his class to illustrate this dilemma. The task in this lesson was: Can x^2+1 equal 0 if x is a real number. In this task, learners can reason empirically by trying out numbers in the expression x^2+1 and noticing that x^2 will always give a number greater than or equal to zero. They can also reason theoretically by drawing on the property that as a perfect square, x^2 will always have to be greater than or equal to zero and therefore x^2+1 will always be positive. In planning the task, Mr. Peters expected predominantly empirical reasoning from learners and hoped that through the class discussion he could build on their empirical reasoning to develop theoretical reasoning.

Learners had worked on the task in pairs and submitted their solutions to Mr. Peters the previous day. Mr. Peters had read these and chosen some to put up on discussion. Mr. Peters' choices and his justifications for these show a very clear sense of working hierarchically with the reasoning in the groups' contributions: from those that had serious errors (which he did not expect), through those who reasoned empirically, to those that provided an adequate mathematical justification.

Most learners had argued that $x^2 + 1$ could not equal 0 because x^2 and 1 are unlike terms and so could not be simplified. They did not think about the possibility that if x was given a value, then $x^2 + 1$ could be simplified and would have a value. Only a few groups had taken the empirical route. One of these responses, that of Jerome and Precious, was central in Mr. Peters' plan for moving through the learners' partial contributions towards an acceptable mathematical argument. Jerome and Precious had substituted values for x before answering the question; however, they had not gone further to provide an explanation. In his interviews, Mr. Peters articulated three purposes for his focus on their response: to develop a sense of meaning for the expression as a variable expression; to develop a method of systematically substituting values for x; and to show how Jerome and Precious' strategy could be taken further to a more general explanation. Mr. Peters had written Jerome and Precious' response on the board and in the two extracts below we see him pressing Jerome in order to achieve these purposes (the presses are in bold font):

121	Mr. Peters:	Let's look what Jerome and Precious did, let's just go and see what Jerome and Precious did, so Jerome and Precious, I hope you gonna explain what you did here, and Lebo and Boitumelo and Tebogo and Marlene also did something similar, ne, **Jerome, what were you thinking?** Come let's see, come let's give Jerome a chance
122	Jerome:	Sir, I said zero equals to anyone of them
123	Mr. Peters:	What, what, what were you doing there Jerome?
124	Jerome:	I said x was equal to two, sir
125	Mr. Peters:	Here, you said x is equal to *(pause)*
126	Jerome:	two
127	Mr. Peters:	x is equal to two, **Why did you choose two, why didn't you choose any other number?**
128	Jerome:	I just took any number
129	Mr. Peters:	You took any number, okay, and what you found out
130	Jerome:	Sir, and I said two times two
131	Mr. Peters:	So what did you do, you took the two into x's place, remember, it's x times x plus *(pause)*
132	Learners:	one
133	Mr. Peters:	Plus one, So you said two times two plus one, And you got an answer of
134	Learners:	five
135	Mr. Peters:	Answer of five, Do y'all see what Jerome was doing
167	Mr. Peters:	Now, I want to know **why did you choose the two specifically, what other numbers did you choose there Jerome**
168	Jerome:	I said x is equal to nought, sir
169	Mr. Peters:	You said x is equal to zero
170	Precious:	Yes, sir
171	Mr. Peters:	**Why did you choose zero, first you chose two, now you're choosing zero, why**
172	Jerome:	I chose any number sir
173	Mr. Peters:	I want to believe you are choosing it for a reason and not just taking any number
174	Jerome:	I was trying to get out, to see if it ends up on nought, sir

In the first extract, Mr. Peters used two presses for meaning (lines 121 and 123) and one press for justification (line 127), where he pressed Jerome on why he started his substitutions with the number two, trying to press Jerome to articulate his trial and error strategy for the class. Jerome's responses to these presses were short and did not elaborate his thinking. He stated that the number two was not particularly significant, a response that Mr. Peters accepted temporarily in order to work with the class on the process of substituting the two and other numbers into the expression.

In the second extract, Mr. Peters pressed much more specifically on Jerome and Precious' choice of numbers, focusing again on their choice of two to start with and their subsequent choice of zero (lines 167 and 171). Mr. Peters was still trying to support Jerome to articulate their systematic trial and error strategy. This was particularly important since so few other learners even thought to attempt an empirical approach. Again, Jerome argued that their choice of numbers was not systematic, and it was only after Mr. Peters inserted his own opinion that Jerome and Precious were thinking systematically and did have reasons behind their actions, that Jerome began to articulate his reasoning – he was trying to get the expression to be zero.

There are a number of points here that are important to my argument. In an interview about this exchange, Mr. Peters said that he had chosen to spend some time reflecting on Jerome and Precious' solution, both because it reflected reasoning that others could learn from, a systematic trial and error strategy, and because the reasoning was limited and needed to be supported towards a more theoretical justification. He wanted to try to build an understanding of why a theoretical justification was necessary. He was pressing on Jerome's thinking to make it accessible to the class (and Jerome) so that they could consider the strengths and weaknesses of the argument. However, in doing this, he was faced with a dilemma in relation to how he interpreted Jerome's responses.

Jerome's responses to Mr. Peters' presses were short replies that did not elaborate his thinking. Even after pressing for an extended period of time, Mr. Peters struggled to support Jerome to articulate his strategy. It was only after inserting his own opinion that Jerome did have a reasoned strategy (line 173), possibly removing some of the pressure from him, that Jerome was able to articulate at least some of his strategy. Mr. Peters' press moves may have been unsuccessful initially because Jerome interpreted them as low presses (Kazemi and Stipek 2001) not requiring a conceptual justification, or as repeated questions signalling that his responses were not correct (Edwards and Mercer 1987), rather than as Mr. Peters intended, that they could provide helpful steps towards a solution.

In an interview, Mr. Peters articulated another possibility, which reflected part of his dilemma. He knew Jerome as a shy learner who may not have appreciated the extended attention on his reasoning, even if it was correct. At the same time, Mr. Peters believed that he should persevere, believing that Jerome's reasoning was helpful and that he could articulate his reasoning for the benefit of both the class and himself. So, even if Jerome had understood the norm that Mr. Peters was working with, that repeated questions mean an attempt to understand his thinking rather than achieve a correct answer, he still may not have been comfortable in expressing his thinking, an interpretation supported by the terse way in which he finally did so (line 174). In this case, articulating the strategy seemed to require a more authoritative re-voicing (O'Connor and Michaels 1996), which Mr. Peters did choose (line 173), to support Jerome to see the strength of his thinking[1].

So in this case, pressing was not immediately useful in supporting Jerome to elaborate his thinking and created a dilemma for the teacher, which he managed by partially removing the pressure from Jerome through using his own authority. Mr. Peters also articulated a second aspect of this dilemma: focusing in detail on one learner's approach at the expense of other learners. Mr. Peters' learners had weak mathematical knowledge, and he worried much of the time about getting through the work. At the same time, he was clear about his teaching goals: he wanted to teach both an important mathematical concept, that the expression $x^2 + 1$ will take a range of values depending on the value of x, and a mathematical practice (Ball 2003) that there is a systematic, justifiable process that can be followed in trying to find out and justify whether $x^2 + 1$ can equal zero[2]. According to Mr. Peters:

They could not understand x^2 plus one to be a unit value if x is any real value ... Teachers in the lower grades place too much emphasis on x^2 plus one having two terms and don't emphasize its unit status if the x-value is known.

In trying to shift a misconception, give meaning to the expression and teach the systematic nature of testing conjectures, which the misconception had prevented many learners from even attempting, Mr. Peters was again working with the dilemma of connecting learners' partial ideas with appropriate mathematical knowledge. He had previously focused directly on the misconception by asking learners to comment on it, but the learners did not manage to see $x^2 + 1$ as taking variable values. So, he took the time to work with Jerome's thinking to help make this point. However, focusing on Jerome's reasoning meant that other learners were not brought into the conversation to try to link the new ideas with their own. Mr. Peters did ask other learners to participate procedurally in substituting numbers into the expression and obtaining answers but did not ask them to comment on Jerome's strategy because it had taken so long for him to articulate. After the above extracts, Mr. Peters articulated Jerome's strategy clearly for the class, which led to a consideration of $x = -1$ and the development of the theoretical argument. However, there is no evidence that most of the learners connected this process with their original misconception or that they shifted this misconception.

The above analysis shows Mr. Peters working with mathematical knowledge at a number of levels. He managed to co-ordinate a range of learner ideas, which demonstrate qualitatively different levels of reasoning, as well as mathematical concepts and practices that he wanted to teach. The fact that he could work simultaneously with these layers of knowledge produced a key dilemma for him: whether or not to press on a learner's idea. In making a strategic choice (Boaler and Humphreys 2005; Lampert 2001) to press on one learner's reasoning to try to

---

[1] Mr. Peters' statement in line 173 can also be seen as "assigning competence" (Cohen 1994) to Jerome, trying to increase his confidence that his method was reasonable and useful, so that he would feel comfortable to articulate it.

[2] I remind the reader that very few learners understood this. Most had argued that $x^2 + 1$ could not be simplified because they are unlike terms.

develop it, he ignored others' – not in the sense that he did not know about their ideas or choose to deal with them (he had earlier) – but because he did not find ways to manage to bring together the different learner understandings within a limited time period. While he found an appropriate way to build mathematical justification from Jerome's original approach, he could not do this for the others in the class. Mr. Peters was painfully aware of this and even after extensive reflection did not think he could have found a way to deal with the situation differently.

To Take Up or Ignore Learners' Contributions?

The range of mathematical ideas that were present in Mr. Peters' class was relatively constrained by the task and possibly by Mr. Peters' extended focus on individual pairs' contributions and thinking. In Mr. Daniels' class, the tasks and Mr. Daniels' pedagogy created a situation of many different contributions, not all of which could be given time in the classroom, creating a dilemma of whether to take up or ignore learners' contributions. While Mr. Daniels' also pressed learners ideas, he did it in different ways from Mr. Peters, with different consequences both for him and for the learners. Mr. Daniels' task asked learners to explore what happens to the turning point and the equation when the graph $y = x^2$ is shifted horizontally on the x-axis. The transcript below is part of a group's report on the question: what happens when you shift the graph $y = x^2$ four units to the left (Mr. Daniels' press moves are in bold).

50	David:	If you move it four to the left, then all your x-values will decrease by four, and the, your y-values again will stay the same, your new equation would be y equals x squared minus four
51	Mr. Daniels:	(Writes $y = x - 4$ on the board)
52	David:	And all, all your, all your coordinates will sit in the second quadrant, therefore, all negative, all your x-values, sorry, will all be negative, and there's again, another pattern, it still increases by one, so it will go negative seven, negative six, negative five, negative four, negative three, negative two, negative one
53	Mr. Daniels:	**David, do you mind just, I just want to see what you mean by that, that there, there's a pattern.**
54	Learner:	Yes.
55	David:	There's a pattern.
56	Mr. Daniels:	**Okay, do you mind just drawing the table on that side (points to board to the right of class), I just want to see**

I discussed this extract in Chap. 9, as an example of Mr, Daniels' responses to appropriate errors and missing information contributions, arguing that when these occurred simultaneously, Mr. Daniels' chose to focus on the missing information contributions. Here, I focus on some of the reasons for these choices, which illuminate the dilemma. Faced with a range of ideas, Mr. Daniels noted two aspects of David's report: the (incorrect) equation, which he wrote on the board as David was speaking; and the pattern, which he pressed David to elaborate on (lines 53 and 56).

As noted earlier, these were presses for meaning, and were authentic. Since the group had not put up enough examples, it was difficult to see where they saw a pattern. David wrote up the pattern but did not respond further to these presses and Mr. Daniels did not press further but moved on to another group. Both the pattern and the equation were on the board and these were referred to in later discussions. However, the claims that all the points were in the second quadrant and that all the x-values were negative were not at all taken up in the discussion.

Mr. Daniels' responses to this and other groups suggest that his main goal in this part of the lesson was to get a number of ideas into the public domain, to be discussed in more detail at a later point. He pressed on an idea that he was not clear about, suggesting that he wanted to understand the learner's thinking. The other ideas, even though not correct, did not receive any attention (except for writing up the equation), possibly because he wanted other learners to respond to them later. So, Mr. Daniels made a choice to get many learner contributions into the discussion, rather than to try to develop particular contributions towards his mathematical goals. This necessitated ignoring some contributions and clarifying others, and he focused on a range of contributions rather than bringing the contributions into relationships with each other.

Later in the lesson, Mr. Daniels invited learners to respond to each other's report backs. At one point, learners asked why, when the graph shifts four units to the left giving a turning point of $(-4,0)$, the sign in the equation $y=(x+4)^2$ is positive. This began a long discussion, in which learners focused on the superficial features of the graph, until one learner, Winile, supported by Mr. Daniels (see Chap. 9) made the important breakthrough that the graph gives a relationship between x and y that is not obvious and they would need to perform some calculation to find the co-ordinates of the turning point from the equation (see also Chap. 4). Winile explained this idea to the class; however, she used examples in her explanation that were not entirely correct and that some learners found difficult to follow. The teacher asked if others would like to comment on Winile's explanation or "help to make sense of it".

195	Teacher:	Sh, Okay, you want to try, Maria, sh,
196	Melanie:	Sir but is the explanation right?
197	Teacher:	I don't know, what do you think?
198	Melanie:	I dunno that's why I'm asking you
199	Teacher:	That's why I'm asking, that's why I'm asking Maria what she thinks.
200	Melanie:	No but you know sir
201	Teacher:	Maybe she can explain it.
202	Maria:	I'm gonna try. Sir, I think here, on the equation, y equals *(inaudible)*
203	Learners:	*(Talking),* Shh
204	Teacher:	Guys, come on.
205	David:	Sir I disagree with her
206	Teacher:	You disagree with that? Okay I'll give you an opportunity to disagree. Let's have Maria, then we'll have David.
207	Winile:	Sir, can I just tell you something before David answers. We supposed to get the y, aren't we supposed to get the y, what the y equals. We're not supposed to get what x is equal to, we getting what y is equal to. So we supposed to, supposed to substitute x to get y.

208	David:	But you get negative one, negative one multiplied by itself can't give you a negative one
209	Winile:	No, I was making an example, I was making an example,
210	David:	We can just put it *(inaudible)*
211	Learners:	It was an example
212	Winile:	Do you understand what I was doing?
213	Teacher:	I, I like the way you were put it now, Winile, the way you said, the way she said that she's getting *y*, she's not getting *x*, you're saying that, she's getting the *y*-value when she substitute the *x*, so, what, what she's saying is that, that squaring will square, give you the *y*-value after you square it, not the *x*-value, that is what she is saying.
214	David:	Sir, but you can get it another way.
215	Teacher:	Okay, but Maria, you wanted to say something also.
216	Maria:	Ag, sir, I'm confused now. *(Shakes her head, class laughs)*

As Maria tried to explain her interpretation of Winile's points, Melanie and David both commented on Winile's ideas without waiting for Maria to talk. Mr. Daniels refused Melanie's request that he authorize Winile's explanation saying that he hoped Maria would explain, and also asked David to wait so that Maria could have her turn (see Chap. 11 for a more detailed discussion). Winile jumped in to re-make her point, which set off a conversation between herself and David. When Maria finally got to speak, she had forgotten what she wanted to say.

One interpretation of this interaction could be that learners were not obeying norms of turn taking, and the teacher was not managing this process properly. However, we see that each of the teacher's interventions made the point that learners should respond to each other's ideas and do so in turns, and he was desperately trying to give Maria her turn. A second interpretation suggests that the learners were deeply engaged in the conversation and had points to make and questions to ask. This was particularly the case because Winile, while providing a useful explanation, had used examples that did not quite work and had made some mistakes. So, there was a lot to comment on in her explanation. While Winile and David's engagement throughout the lesson might be seen as argumentative, in fact their intellectual antagonism became a source of very productive ideas for the class to work with and increased the engagement of many learners. Mr. Daniels was treading a fine line between supporting individual learners to make their points in ways in which they could be heard and supporting a climate that excited learners and helped them to stay engaged.

In the extract, Mr. Daniels did not make use of many press moves, in fact there was only one (line 197) which he did not follow through because he wanted to give Maria her turn. In choosing to support many learners' contributions, Mr. Daniels chose not to press for too long on individual learners' contributions. Mr. Daniels noticed the significance of many of the learner contributions and tried to give them all a chance to be heard. However, doing this, led to at least one contribution, Maria's, being ignored.

Mr Daniels' dilemma was both different from and similar to Mr. Peters'. Differently from Mr. Peters, he chose to allow a range of contributions and ideas to be stated publicly, which meant that he ignored a number of them and also struggled to develop further those that he did not ignore. Similarly, to Mr. Peters, he had two clear goals for his teaching. First, he wanted to challenge the learners' misconception about the relationship between the turning point and the equation by supporting them to see that the relationship between the graph and the equation was more complicated than they thought. Second, he also wanted them to justify their reasoning both as a means to develop their mathematical thinking and as an end in itself. Similarly, to Mr. Peters, Mr Daniels also struggled to see how to bring the learners' contributions into contact with the concept and practice that he wanted to teach. Differently from Mr. Peters, he chose to allow a range of contributions to help him do this. This approach did work to some extent, because Winile and David did push each other to articulate the relationship.

The analysis of Mr. Daniels' teaching illuminates how, when teachers support learners to express their thinking, they can be faced with a barrage of contributions. Some of these might be related to the original contribution and some might not be, some might be helpful for the mathematical direction of the lesson and some might not be, some might be worth considering in their own right for the new mathematical insights they might support, and some might be so poorly expressed that they are difficult to understand without much more work on the part of the teacher and the class. Reform-oriented teachers want to value, support, and encourage such learner engagement, however, making decisions as to which contributions to take up and which to ignore, and keeping track of the full range of contributions present challenges for teachers. Noticing significance becomes more demanding as teachers need to keep track of: (1) the original question and the extent to which learner contributions do address and answer it; (2) the extent to which learner contributions address each other's points; (3) additional points that emerge in subsequent contributions; (4) errors or misconceptions in the contributions that are made and that might need some discussion and work; and (5) who has spoken and who still needs to speak – a task often complicated by excited interruptions from learners, as seen above. In addition, teachers also have to find ways to help learners to keep track of the discussion. If the teacher finds it demanding to keep track of the relevance of the contributions, it is far more demanding for the learners. In many cases in both Mr. Daniels' and Mr. Peters' lessons, learners could not respond appropriately to previous ideas because they did not understand their significance and so further contributions did not always build on previous ones (see Chap. 11 for a discussion of some of the consequences of this). Managing these situations requires steering a path somewhere between Chazan and Ball's (1999) unproductive agreement and unproductive disagreement. The teachers in this study faced agreement and disagreement, productive and unproductive ideas all at once and had to manage this range while making progress with the mathematics.

Conclusions

In the two scenarios described in this chapter, I have illuminated how two knowledgeable and thoughtful teachers were confronted by dilemmas as they engaged with learners' meanings to develop their mathematical reasoning. That they were confronted with these dilemmas is not surprising – a number of other more experienced and knowledgeable teachers and researchers have written about similar dilemmas. What I have shown is how closely related these dilemmas are to reform pedagogy, particularly the *press* move. Choosing which learner contributions to work with, when to work with them and for how long, in order to build appropriate mathematics is a feature of all the dilemmas described here and in the literature. At the heart of both Mr. Peters' and Mr. Daniels' dilemmas was an attempt to value and engage with learner contributions while simultaneously transforming them into appropriate mathematical concepts and practices. Both were working with the idea of mathematical reasoning as an important part of mathematical proficiency (Kilpatrick et al. 2001) supporting and supported by mathematical concepts and relationships. Both struggled, in different ways, to bring learners' partial meanings and intuitive reasoning and justification into contact with established mathematical ideas.

I have argued that the "press" move, privileged by reform discourse for supporting learners' thinking can have varying success and is implicated in the dilemmas that the teachers experienced. In Mr Peters' class, a series of extended presses needed to be supported with the teacher's authority that the learner had something valuable to contribute, in order to support the learner to articulate some of his reasoning. In Mr. Daniels' class, the demands of managing a range of excited contributions meant that pressing was not always useful. When the teachers did choose to press more extensively on particular contributions, they had to make choices to ignore other contributions. In distinguishing between presses for meaning and presses for justification, I suggest that it might help teachers to know which of those they are using, just as Kazemi and Stipek (2001) distinguish between high and low presses. However, I am also suggesting that continued pressing for learner thinking can be counter-productive, even if learners understand the norms with which the teacher is working. This supports my argument in Chap. 9 that pressing needs to be strategically combined with a range of other moves, including bringing the teacher's authority into the discussion through revoicing (O'Connor and Michaels 1996), conceptual explanations (Lobato et al. 2005), or strategic episodes of closed questioning (Brodie 2007c).

I would like to end this chapter with a quote from Mr. Daniels, made as he reflected on the above incidents in his classroom:

> It is quite demanding to negotiate between learners' understanding, trying to listen to what they do, and at the same time keep track of where you want to go to, and also with negotiating their arguments, listening at the same time while negotiating and keeping track where you want to go, and then **you are bound to slip up … I thought that preparing will help but that is not why it happens.** You can still prepare and you can still be confronted with all these issues because the focus is not just on what you are saying but also you are thinking about what the response is and what just happened before that (my emphasis).

That Mr. Daniels believed he had slipped up when not managing to take account of all learner contributions in his classroom, suggests a long-standing view that teachers need to be perfect, that they do not make mistakes. This is true of the reform literature as well, we present ideal images of what mathematics classrooms should look like, images which may be very difficult to live up to, except perhaps for short periods of time. Even the literature on dilemmas suggests that teachers can manage these in ways that are deemed to be successful. The only times when mistakes can be acknowledged is when teachers are learning reform practices (Heaton 2000). My analysis suggests that teachers who work with reform-oriented pedagogies and seriously grapple with the demands that arise, will both be successful and unsuccessful: they will at times choose appropriate moves which help some learners to articulate their reasoning; and while taking up some ideas, they may ignore other potentially fruitful lines of conversation, either consciously or because they did not notice the significance of particular contributions. Teachers' more and less successful moments are not steps along a road to perfection, but are a part of this way of teaching. Although, as Mr. Daniels hoped, teachers can become more prepared for particular dilemmas, as he found out, there will always be surprises and new challenges as new groups of learners interact in new ways around mathematical ideas.

Chapter 11
Learner Resistance to Teacher Change

The previous chapters have described a range of ways in which five teachers shifted their pedagogies to take account of learner contributions and to teach mathematical reasoning. The analysis has pointed to both successes and challenges in their teaching and has shown that a new curriculum and new pedagogies place new and increased demands on teachers and learners. Increased demands can lead to resistance on the part of both teachers and learners. In this chapter, I focus on learners' resistance to their teacher's changing pedagogy.

The phenomenon of learner resistance is often spoken about by teachers but has not received significant attention in the mathematics education literature. Extended resistance in Mr. Daniels' classroom suggested that this was an important area to explore in this study. As Mr. Daniels shifted his pedagogy to include learners' mathematical reasoning, so the learners began to resist this shift in demands on them, at first subtly and then openly. In this chapter, I explore the dynamics of this resistance, including both the learners' and teacher's contributions to it. I argue that it is not possible to avoid learner resistance to changing pedagogy, and that in fact, a certain amount of resistance is useful and positive. However, too much resistance can be destructive and teachers can be aware of the potential for resistance and the possibilities of managing it.

Resistance to Pedagogy

Research into resistance in schools has identified a range of complex dynamics that come together to produce resistance. In his seminal study into the cultural reproduction of schooling, Willis (1977) showed how working class boys explicitly rejected the hegemony of middle-class schools and society through the creation of counter-cultures. However, in doing this, these boys further entrenched their marginalized positions because they did not gain access to the very resources that could shift their material conditions, the material benefits of schooling. So resistance both requires and creates collusion with the very structures that are being resisted. McNeil (1981, 1999) further illuminates this collusion among teachers and students and shows how it creates the phenomenon of "defensive teaching". Her study

K. Brodie, *Teaching Mathematical Reasoning in Secondary School Classrooms,*
DOI 10.1007/978-0-387-09742-8_11, © Springer Science+Business Media, LLC 2010

showed that teachers who had deep knowledge and convictions about issues in the social studies curriculum, nevertheless, simplified and fragmented what they taught in class in order to avoid controversy in their lessons. Students responded to this simplification by acquiescing to classroom procedures: taking notes, not making contributions or asking questions, and publicly accepting the "facts" that the teachers presented. However, interviews with the students showed that they often did not believe what their teachers told them, particularly when they had access to other information, or held strong views on particular issues. They silently resisted the knowledge offered in their classrooms, learning what was required to pass tests and keeping quiet about their own opinions. McNeil argues that defensive teaching stems from teachers' fears that allowing a variety of opinions and discussions around controversial issues might create a loss of control in the classroom. Students also preferred not to risk their own ideas and beliefs in public discussion and so publicly acquiesced to the teachers' control while privately resisting it. Thus, "teachers and students contributed to each other's (and their own) alienation in the classroom which further reinforced tighter knowledge control and increased resistance to it" (1981, p. 323). In the South African context, Chick (1996) argues that teachers' and learners' resistance to communicative language teaching can be explained by the "safe-talk", that they engage in; highly stylised interaction patterns with substantial chorusing by learners. He argues that "safe-talk" represents collusion by teachers and learners to maintain dignity in the face of the "unpleasant realities" of the South African education system, which deny real access to knowledge to so many. Becoming more communicative would remove this defensive strategy, hence the reason why teachers and learners resist any significant shifts in interactional style.

The above research helps understand the seeds of resistance to new pedagogies. Because collusion and defensiveness bring some form of safety in traditional modes of interaction, both teacher and learner will resist shifting these, even as these represent forms of resistance to current structures of schooling. One key vision of reform pedagogies is to overcome traditional authority relations between learners and the teacher, where the teacher is assumed to be the sole authority in the classroom, particularly in relation to knowledge. The above research argues that such authority relationships both produce and are produced by resistance. In reform pedagogies, learners are encouraged to both author and authorize mathematical knowledge in the classroom, through expressing their own reasoning and evaluating the ideas and reasoning of others, thus overcoming some of the reasons for resistance in traditional schools. In such classrooms, the epistemic authority derives from the mathematics itself, which provides the grounds for reasoning and justification (Ball and Bass 2003; Kilpatrick et al. 2001). Resistance to reform pedagogies, thus, presents a paradox for teachers and researchers. Why do pedagogies that attempt to overcome resistance in fact continue to produce resistance?

Research into learner resistance to reform pedagogy suggests a number of possible answers to this question. Particularly, but not only, in developing countries, the authoritarian nature of many education systems ignores teachers' perspectives and new curricula, and pedagogies often do not find points of contact with teachers' realities in order to provide appropriate guidance in relation to the ideas of reform (O'Sullivan 2004; Tabulawa 1998; Tatto 1999). Reform pedagogies often clash with deep-rooted beliefs about knowledge, learning and the purposes of schooling and with teachers', learners' and parents' images of classrooms as places where the teacher is in control of the knowledge produced (Association for the Development of Education in Africa (ADEA) 2003; Cuban 1993; Shamim 1996; Tabulawa 1998). However, as Shamim (1996) notes, this is not enough of an explanation, and it poses a serious paradox for teachers and researchers. If learners, teachers, and parents expect compliance to authority, then why is this compliance "functional only in terms of the social organization and shared norms of interaction of the traditional classroom" (p. 118). Even though new pedagogies encourage a shift in authority relations, they are still underlined by a notion of authority, and it is still the teacher who, from a position of authority, is introducing the changes. Thus, further explanation is required.

Reform approaches require more work from both teachers and learners; defensive strategies are relinquished, and learners are expected to articulate their thinking for the consideration by the teacher and other learners. As Chick (1996) notes, these approaches are more threatening in two ways: learners have to express and take responsibility for their ideas publicly and the teacher has to feel competent to respond appropriately to learners' unexpected ideas (see Chap. 10). Reform approaches also require that the teacher and learners become comfortable with some loss of certainty in the classroom, ideas are discussed and debated, and the "right" answer is not always forthcoming or easy to discern (Copes 1982; Kloss 1994; Lubienski 2000). It is clear that reform teaching demands more work and more risk from both teachers and learners, adding to the challenges in the classroom and opening possibilities for resistance.

Much of the research into resistance to reform pedagogy views resistance as problematic for developing new ways of teaching and proposes ways of overcoming resistance. Cobb and his colleagues' work on classroom norms argues that learners can be taught different ways of interacting in classrooms, which is necessary for new pedagogies to be enacted successfully (McClain and Cobb 2001). In traditional classrooms, teachers explain ideas, ask questions to which they expect correct answers, and strongly evaluate learners' responses as correct or incorrect. In reform pedagogy, appropriate norms for interaction are that mathematical reasoning, justification, and communication are as important as correct answers and count as appropriate contributions; partially correct or incorrect contributions are helpful and often illuminate important ideas; and learners are expected to listen to and comment on their classmates' contributions in respectful ways. Teachers who teach mathematical reasoning do continuous work on developing these norms (Boaler and Humphreys 2005; Lampert 2001; Staples 2004). Boaler (2002) argues that students need to be taught how to participate in reform classrooms and that teachers should explicitly develop learning practices among learners such as how to begin exploring open-ended problems and how to explain and justify ideas. In the Namibian context, O'Sullivan (2004) has developed a programme of action research with teachers where they actively challenge assumptions about both traditional and reform pedagogies. While developing ways of overcoming learner resistance are important and extremely helpful for teachers and teacher-educators trying to develop the teaching of mathematical reasoning, they can lead to a dangerous assumption: that all forms of resistance can and should be overcome and if they are overcome, teaching can proceed smoothly. This represents an approach to teaching and knowledge about teaching that suggests that problems of teaching can be solved rather than managed, a position eloquently argued against by Lampert (1985) (see Chap. 10). In this chapter, I suggest that while openly expressed resistance on the part of learners might be uncomfortable for teachers who are already engaged in the difficult work of developing new pedagogies, such resistance can be a sign of healthy interaction in the classroom. At the same time, I suggest that teachers can consider the particular dynamics of resistance in their classrooms and find ways to work with it.

One of the key claims of reform pedagogy is that it responds to theories of learning, particularly constructivist and situated perspectives on learning. Constructivist perspectives argue that learners construct new knowledge through engaging with their social and physical environments (Hatano 1996). Different environments provide different possibilities and constraints for the construction of mathematical knowledge (Hatano 1996). Situated perspectives argue that learning occurs as learners become better participants in communities of practice, using the intellectual resources and tools of these communities. Both of these perspectives argue that resistance is an inherent and important part of learning. From a Piagetian constructivist position, resistance is inherently part of the processes of assimilation and accommodation which together create the process of equilibration (Rowell 1989). To develop new knowledge from this perspective it requires that the learner gives up a current equilibrium and takes on the difficult task of restructuring and reorganizing her knowledge (Hatano 1996). All learners resist this task, trying successive compensations in order to avoid real transformations in knowledge organization. It is only when the integrity of our current knowledge is so seriously threatened that the task of reorganizing knowledge becomes a preferred process. From situated perspectives, resistance and contestation take place in communities of practice (Wenger 1998). Lave and Wenger (1999) argue that the process of legitimate peripheral participation is one that must involve resistance, from both newcomers and oldtimers. In order for practices to grow, newcomers must both accept and challenge the practice; they need to both engage existing practice and within this assert new ways of knowing and being. Oldtimers may resist challenges to the practice, wanting to preserve their preferred ways of action; however, they know that the practice must grow and change in order to survive. In the case of classroom reform, particularly in secondary schools, learners are oldtimers, having participated in traditional classrooms for many years. In a strange twist of roles, teachers introducing new ideas may take on aspects of the newcomer's role, introducing new ideas into practice, which learners may resist. Practices are always sensitively balanced in relation to new ideas, because they need to be open to these in order to grow and yet maintain a core stability and regularity to continue the practice. Thus, contradiction and resistance are key to growing practice.

In what follows, I show how this balance both shifts and is maintained in Mr. Daniels' classroom. The dynamics of the resistance that he encountered are complex. Learners did not consistently resist, and not all learners resisted. Many of the key resistors were also key participants in the conversation, expressing their own reasoning and engaging with the ideas of others. However through the lesson, we begin to see mounting resistance as the ideas in the conversation became more difficult for the learners to access. I show how both learners and the teacher contribute to the dynamics of the resistance and argue that while such resistance can be seen as positive, indicating important shifts in practice, it is clearly uncomfortable for both teacher and learners. Resistance can be seen as both positive and negative, and it is not and should not be entirely avoidable. Rather, teachers can develop under-standings about when and why resistance might occur and find ways to manage it.

The Context of the Resistance

The episodes of resistance occurred as learners in Mr, Daniels' classroom were working on the question: What changes as the graph of $y=x^2$ shifts 3 units to the right to become $y=(x-3)^2$, and 4 units to the left to become $y=(x+4)^2$. In Chap. 4, we showed how, through the week, one learner, Winile's, learning developed, informed by an extended conversation. Here, I show how, later in the week, open resistance among the same classmates to the same conversation influenced the course of the lesson and subsequent learning. The conversation was in response to a question established by both teacher and learners as an important question to dis-cuss: Why does a negative sign in brackets correspond to a shift to the right and a positive turning point; and a positive sign correspond to a shift to the left and a nega-tive turning point? This question was co-produced in conversation between two learners, Michelle and Lorrayne in relation to a report-back from Winile's group.

Mr. Daniels worked hard to facilitate discussion around this question. First, he tried to establish a common ground (Staples 2007) making sure that all learners understood and were interested in the question. The following extract shows how he spent substantial time in doing this.

100	Michelle:	And then if you look at y equal x plus four, why is it that the turning point is a negative.
101	Learner:	But the equation is positive
102	Michelle:	And the drawing is positive.
103	Learner:	I asked that too. *(Some learners laugh).*
104	Learner:	I'm also asking the same question.
105	Mr. Daniels:	What question are you asking?
106	Michelle:	The question …
107	Mr. Daniels:	Yes.
108	Michelle:	Look at our drawing where …
109	Mr. Daniels:	Okay. Where's my drawings? *(finds drawings)*

(continued)

110	Michelle:	Where it says y equals x plus four on the left hand side.
111	Mr. Daniels:	Right.
112	Michelle:	Our turning point is a negative four.
113	Mr. Daniels:	Okay
114	Michelle:	Then Lorrayne that said with the one on the right, where it says y equals x negative three squared, and the turning point is a positive. Because you squaring it, it will become a positive. But what happens with um, the one on the left?
115	Lorrayne:	The negative one. The equation is positive but the graph is on the negative side.
116	Mr. Daniels:	The equation is positive but the graph is on the
117	Learners:	Negative side
118	Mr. Daniels:	Okay David. Do you know how to explain that? Do you have any ideas on that?
119	David:	Sir, you people are taking my graphs and making other things with them.
120	Mr. Daniels:	Okay. Now that's why I'm specifically asking you because I know that you sitting there with something that you've got an understanding of. And we want you to share it with us.
121	David:	No, I got that graph. Is that what you want
122	Mr. Daniels:	No, no no. That question is … That's why it's important for you guys to listen. We need to learn to listen to each other and talk to each other to help each other to learn. That's the point of being in this class in the first place.
123	David:	Sir, can you please say the question again.
124	Mr. Daniels:	The question is, they asking, if you look at the equation ok, the graph there, the one on the left hand side. They say that the turning point is minus four and the equation is y equal x plus four all squared. Why do I have a negative turning point there?
125	Mr. Daniels:	Am I interpreting your question correctly?
126	Learners:	Yes.
127	Mr. Daniels:	Is that what everybody is struggling with?
128	Learners:	Yes Sir.
129	Mr. Daniels:	What do you think David?
130	David:	Sir they saying, they asking you, why do you have your, why's it a negative turning point when it's a positive?
131	Mr. Daniels:	It's minus four is the turning point but it's a plus four inside in your equation.

In lines 100–104 Michelle repeated part of the question that she had asked previously, relating to the graph $y = (x + 4)^2$. A classmate helped her to re-articulate the question, while another confirmed that she too asked the same question. Mr. Daniels built on this asking for yet another re-articulation of the question in lines 105–108. His move to "find his drawings" in line 109 signalled the importance of using the graphs to understand the question. In lines 109–117, Michelle, helped by Lorrayne, re-articulated the question slowly, for all to hear. Mr. Daniels helped to slow down the discourse by coming in every other line to affirm the statement of part of the question and by repeating the key part of the question in line 117. Mr. Daniels then asked David to try to answer the question, which he did not do, and asked for the question to be repeated, which Mr. Daniels did in line 124. He then checked that he had in fact stated the question correctly (line 125) and reconfirmed that this was the question that the class was struggling with. By the end of the extract above, the question had been stated and re-stated a number of times and a number of learners had indicated interest in the question, suggesting that there was a common ground (Staples 2007) on which to build.

In the above extract we see Mr. Daniels doing some "norm work", first in articulating the question as an important, shared question and second in directly communicating key norms about the discourse (lines 120 and 122). He stated his opinion that David had something useful to share, thus assigning competence to David (Cohen 1994) and indicating an expectation that learners with knowledge should share it with the class. He stated that it was everyone's responsibility to listen to each other, so that they could help each other with the question, and also acknowledged the girls as authors of the question by asking if he was interpreting the question correctly (line 125). Such "norm work" was common in Mr. Daniel's classroom and always worked towards his goal of enabling learners to collaborate to achieve understanding (see also Chap. 4).

After the above extract, an extended conversation about this question ensued for over 30 min. During the conversation, many learners made extensive contributions, articulated their ideas and developed their thinking substantially. Two learners, Winile and David contributed a number of ideas in response to each other, which were important in constructing both the development of the mathematical ideas through the conversation and some of the resistance.

In her initial presentation on behalf of her group, Winile had presented a rather confused argument, suggesting that she did not understand the relationship between the equations and the graphs. Michelle and Lorrayne had picked up on this in their question, which attempted to get at the relationship. After listening to the conversation for some time, Winile made an important contribution:

> the plus four is not like the x, um, the x, like, the number, you know the x (showing x-axis with hand), it's not the x, it's another number … you substitute this with a number, isn't it, like you go, whatever, then it gives you an answer … we're not supposed to get what x is equal to, we getting what y is equal to, so we supposed to, supposed to substitute x to get y

Here, she argued that they could not make a direct link from a superficial feature of the equation, the +4, to a feature of the graph, the turning point. Rather, they had to take account of the underlying relationships between the variables that gave rise to the graph, and that the equation transformed the numbers in ways that were not obvious. This suggests a deepening of her thinking about equations and graphs; she had differentiated the direct equation–graph relationship and now saw a more complex relationship, between the variables in relation to each other and in relation to the graph (see Chap. 4 for an account of how this contribution both emerged from and contributed to preceding and subsequent contributions in Winile's learning trajectory).

Later in the conversation, David argued the following about the relationship between the equation and the turning point:

> Okay, every one agrees that your standard equation we got first was this, right [writes $y = x^2$]. That was the first one we got. Another one we got is y equals x, no wait, plus four

squared right *[writes $y=(x+4)^2$]*. Now that's the one we got. Now if you want to find out where to put your graph, you take out your constant, alright, just put it over there, then you'll get, you, ja, your y equals x squared back, that's your standard graph, you take this constant, you put it over there *[indicates putting it on LHS]*, okay, which will give you negative four, okay, plus y, x squared, and then now because it's negative four, you'll have your graph, one, two, three, four, and then you need to know, you need to, your turning point will be there, because it's negative.

And a little later he extend the argument to the case of $y=(x-3)^2$ giving a turning point of positive three.

> Yes, I mean, look if this was a negative, okay, let's take another one, x minus three here, squared, take that out, it gives negative three, okay, then y, then your three will become a positive, you see, because you're putting it on that side of the equation, then you need to put it three spaces forward.

David's argument combined a form of inappropriate algebraic manipulation "take out your constant, alright, just put it over there" with some of the exploration they had done previously in shifting the graphs. He did not articulate his argument very clearly, but he could have been arguing that $y=(x+4)^2$ is the same as $y=x^2+4$ and so the four could be "taken over to the other side" and become a negative four.

In the extract below, Winile challenged David's argument, saying that he was telling them what to do, rather than explaining why, which is what Michelle had asked. She also did not articulate her challenge clearly. Her previous contributions suggested that she was looking for a relationship between the x and y values in the equation and the graph, but she claimed that David was telling them about the x-value rather than the y-value.

279	Winile:	*(Laughs)* Okay sir, he's just telling us where to put like, the turning point of the graph, and we want to know why, the y-value is, we want to know what the y-value is and you're telling us the x-value.
280	David:	That would give you a y-value of nought.
281	Winile:	You gave us four plus four,
282	Mr. Daniels:	Just say that again, what did you say,
283	David:	That would give you the y-value at nought, each one of those.
284	Mr. Daniels:	Just come explain that again.
285	David:	*(Shakes his head)*
286	Learners:	*(Talk simultaneously)*
287	Mr. Daniels:	You said now that will give you the y-value of nought, now explain, how does it give you that.

The way in which Winile articulated her challenge allowed David to respond that the y-value is 0, without having to explain how this answered the question. Mr, Daniels realized that if David could make the connection between the y-value of 0, which he had merely stated, and the x-value, he would make some progress in answering the question, so he pressed David to explain. Mr. Daniels asked David three times to explain how he got the y-value of 0, but David refused. It was at this point that two learners, who had previously both contributed to the conversation and expressed some resistance to it, erupted into a full-blown challenge to the conversation.

Learner Resistance

Although much learning took place through the conversation, and many learners openly expressed enjoyment of the interaction and their learning, a point came in the lesson where some learners began to openly express resistance to the conversation. An analysis of this resistance from the perspectives of the learners and the teacher illuminates some of the dynamics of resistance and suggests that learners' resistance is not a unitary construct. Different learners will resist for different reasons.[1] In this case, I show that the two main "resistors" in the lesson, Michelle and Melanie, showed both similarities and differences in their orientations to the conversation, their learning and their resistance to the conversation. I argue that their resistance relates to their different orientations to learning and teaching and sets up different dynamics for the teacher to deal with.

The following extract follows immediately after the previous one.

288	Michelle:	If you had the chance you would keep us here until tomorrow
289	Mr. Daniels:	I will keep you here until tomorrow
290	Learners:	Aah, sir (*talk simultaneously*) (*Michelle sits back and folds her arms, other learners, including Melanie have their hands up*)
291	Mr. Daniels:	Because, because, ai, this is what I want, I want you people to talk to each other, I want you to talk about the mathematics and it's happening, so I'm happy about this, very
292	Melanie:	But when I go home what did I learn (*laughter*)
293	Mr. Daniels:	You'd be surprised
294	Melanie:	Excuse me, sir, I, okay, sir, you can see we are so battling with this but, you refuse to just give us the answers
295	Learner:	Because we just get more confused
296	Learner:	Ja, but don't give me that
297	Michelle:	How long are we going to play this game (*learners talk simultaneously*) man, I can play the game until it's over
298	Mr. Daniels:	Sh, guys, listen up, (*inaudible*), I think I need to say something here, this is not a game, okay, this is mathematics that you need to understand, okay, there's, there is mathematics that you need to understand, there is mathematics that is in your syllabus and you need to make sense of it, I just don't want to give you answers, which means nothing to you, If I give you the answer what will it mean to you, really, without an understanding of why the answer is that, and that is the point here, and I like, Winile is making sense of it, David is trying to make sense here, we've got Lorrayne here that contributed, Maria, Ntabiseng had wanted to say something, (*Melanie indicates herself*), you, exactly, that's what I want. So let's, let's
299	Michelle:	Some of us aren't that intelligent as these guys
300	Mr. Daniels:	You don't need to be
301	Michelle:	This guy that comes with the answer out of the blue, I mean, I never did, look at that, I can't just sit here fiddling, and I don't even understand what's going on, no man

[1]McKinney (2004, 2005) explores how different kinds of resistance are related to different kinds of identity work by university students in the context of a critical literacy approach.

Michelle and Melanie had both contributed to the discussion, and were engaged in trying to make sense of the ideas throughout the lesson. Michelle had asked the original question that began the conversation. When Mr. Daniels listed a number of contributors (line 298), Melanie eagerly indicated that she too was one and Mr. Daniels affirmed this.

Melanie's contributions in the above extract suggest that she felt she was not learning anything through the conversation and that the teacher could see that the class was struggling and was refusing to help them (lines 292 and 294). In the next section, we will see Melanie demanding to know whether David's explanation was correct and claiming again that Mr. Daniels was concealing his knowledge from them. Previously she had asked David if he was sure about his contribution, and whether it would "always be like this, forever and ever amen". Melanie's expressed resistance suggests that she was focused on finding correct answers, rather than on trying to understand her classmates' mathematical reasoning. Using the new norms of conversation, she first constituted David as the authority on his explanation, asking if it would always be correct. She initially seemed satisfied with his assertion that it would be. However, as the discussion continued and Winile challenged David's explanation, Melanie constituted the teacher as the authority, as in traditional classrooms, demanding the correct answer. We can characterize Melanie's resistance as coming from a learner who was willing to engage with the mathematics in the classroom but who did not want to author her own understanding, she was more comfortable with the teacher or another learner as the authority. So Melanie was willing to engage with some of the demands of reform pedagogy, to make contributions, to listen, and to evaluate those of others, and to try to come to a resolution. When there was no clear resolution, she demanded one from the teacher.

In contrast, Michelle seemed more interested in authoring her own understanding of the mathematics. She had asked the question that began the discussion and was clearly engaged in trying to work out a justified answer. In the episode above, she was frustrated that an answer "came just out of the blue" and that she could not understand it (line 301). In an earlier exchange, when a classmate, Candy, had suggested as an answer to her question: "can't it just be like that", she was unwilling to accept the response without a proper justification. Mr. Daniels affirmed her demand for a better explanation.

138	Candy:	Sir, couldn't it just be like a basic thing, that if it's on the positive side then your equation is negative and if it's on the negative side then your equation is positive? Can't it just be like that (laughs)
139	Michelle:	I can't accept that
140	Learners:	Mutter, talk over each other
141	Mr. Daniels:	Okay. Let's …. Say that again.
142	Michelle:	I can't just accept that.
143	Mr. Daniels:	So, I'm not expecting you to accept it.
144	Michelle:	No, I'm just saying that I can't …
145	Teacher:	That's good. That's what I'm saying. I'm saying it's good that you don't just accept it

Michelle's references to time in the first extract above: "you would keep us here until tomorrow" (line 288) and "how long are we going to play this game" (line 297) suggests that it was not the struggle to make sense that bothered her, but the fact that it was taking so long. She was concerned that others, who were more "intelligent" than her (line 299), were understanding, but that she wasn't getting it. She also referred to the conversation as a "game" (line 297), possibly building on Melanie's accusation that the teacher was playing with them because he refused to help to resolve their difficulties. Unlike Melanie, Michelle did not demand that the teacher actually resolve the situation, but she did challenge the fact that the teacher seemed willing to allow the conversation to continue indefinitely without giving any sense about what was going on that she could understand.

The above analysis works with the two girls' explicitly expressed resistance. However, it is likely that there were other issues that were bothering them that they could not fully express, although their comments give us a clue. It may have been the case that either Michelle or both girls doubted David's explanation, which was incorrect. Their struggle to make sense of complex ideas was not helped by the barely coherent and sometimes incorrect responses of their peers. Although Winile had challenged David, her challenge was also poorly expressed and it is likely that the other learners struggled to understand her as well. What we see above is not merely resistance to a new way of interacting in the classroom. Rather it portrays the complex dynamics of the cognitive, social, linguistic, and emotional demands of making sense of new ideas in new ways. The fact that there was such resistance suggests that Mr. Daniels was achieving some of his goals.

It is important to note that many learners did not resist, to the contrary many were enjoying the conversation. After Michelle and Melanie erupted, some learners claimed that they had never enjoyed a lesson so much, precisely because of the discussion and debate it engendered. Mr. Daniels did make use of these learners to help make progress in the lesson, nevertheless he was visibly disturbed by Michelle and Melanie's challenges, as many other teachers would be. I now turn to the ways in which his responses to the resistance helped to support and fuel it further.

The Teacher's Contributions

Mr. Daniels both tried to work against the resistance and also inadvertently contributed to it. In the extracts in the previous section, some of his contributions attempted to make clear what his expectations were and how they differed from what is usually expected in mathematics classrooms; the "norm work" that he did throughout the lessons (lines 143, 145, 291, 298). At the same time Mr. Daniels may have unintentionally undermined some of what he was trying to do by making flippant comments. In response to his remark "I'll keep you here until tomorrow" (line 289), Michelle sat back with folded arms looking angry. It is not clear why Mr. Daniels made this remark, nor the next one: "you'd be surprised" (line 293). His first statement suggests that he may have been trying to reclaim some of his authority in the face of obvious resistance. When teachers are faced with these kinds of challenges, they may fall back on well-worn expressions of authority. Trying to reclaim authority in this way can be counter-productive, as it was in this case. Mr. Daniels' first statement made Michelle even more angry and his second made Melanie more demanding that he provide the answers.

So, teachers falling back on traditional norms, even as they teach new ones, may fuel learner resistance, particularly if the learners have engaged in the new demands of exploring mathematical reasoning. One of the key contributors to the learners' resistance in this lesson was Mr. Daniels' successful enactment of aspects of reform pedagogy. In the first extract in the previous section, we saw Mr. Daniels pressing on David for an explanation. He did not help David to give the explanation, wanting it to come from him. It was this pressing, and David's refusal to explain that angered Michelle and that was most likely the "game" that she was refusing to continue to play. Because Mr. Daniels did manage to hold back his own ideas in order to elicit the learners' thinking and when learners' ideas were unclear, he pressed them to explain, rather than re-articulating the ideas himself, other learners such as Michelle and Melanie felt lost and frustrated. Earlier in the chapter, I showed how Mr. Daniels' approach was very helpful in supporting the learners to articulate and re-articulate the question to be discussed so that all could work from a common ground. However, he was less successful in achieving a common ground and shared understandings with respect to the learners' attempts to answer the question, probably because these were more divergent and less clearly articulated (see also Brodie 2007b).

Yet another aspect of Mr. Daniels' enactment of reform pedagogy probably contributed to the learners' resistance. Because he was trying to get learners to talk to each other and to express and articulate their own ideas, he often claimed not to understand what a learner was saying. Sometimes these claims were genuine; he did not understand the learner's idea and needed more explanation himself. At other times, his claims were not genuine; he feigned ignorance in order to obtain more articulation and justification from learners. The following extract illustrates this.

188	Winile:	You see when you got this, plus three [writes $y = x^2 + 3$], you have to substitute this with the, that with like the one, zero, one two, three [Draws numberline, x axis]. Your turning point is here. You have to substitute this with this negative one here, plus three. Do you understand? This three [circles the 3 in $y = x^2 + 3$] is not, is not part of the, this x, uh, variables. Its the given (inaudible) Get it.
189	Learners:	Oh…(lots of muttered comments)
190	Mr. Daniels:	Okay guys, let's, let's,
200	Winile:	y is equal to that four, that's your turning point, then you get your final answer
201	Learners:	Mutter

202	Mr. Daniels:	Can somebody else try, because I saw a lot of people said, oh-ja, now we see, can I just see, Lorrayne, maybe you can put it another way
203	Learners:	(*General chatter – inaudible*)
204	Mr. Daniels:	Sh, okay, you want to try, Maria, sh
205	Melanie:	Sir but is the explanation right
206	Mr. Daniels:	I don't know, what do you think
207	Melanie:	I dunno, that's why I'm asking you
208	Mr. Daniels:	That's why I'm asking, that's why I'm asking Maria what she thinks
209	Melanie:	No but you know sir
300	Mr. Daniels:	Maybe she can explain it.
301	Maria:	I'm gonna try, sir, I think here, on the equation, y equals (*inaudible*)

After Winile's contribution, a number of learners indicated that they wanted to respond. Mr. Daniels indicated that Maria should talk (line 204) but before she could make a contribution, Melanie jumped in, asking whether Winile's explanation was correct. The teacher reflected Melanie's question back to her, claiming not to know and asking what she thought, trying to press her to make her own evaluations of the mathematical truth of Winile's statement, rather than relying on his authority. Mr. Daniels did know that parts of the explanation were incorrect and later in the lesson he directly challenged Winile on her mistake. It may have been the case that Melanie suspected a problem with Winile's explanation, which was why she asked the question. From her perspective, the teacher knew the correct answer and seemed to be deliberately withholding it. However, other learners did not seem to have the same concerns. In the extract Maria was certainly eager to try to explain Winile's idea, as were other learners (line 202).

The above analysis raises some important questions. Was it the case that Michelle and Melanie resisted Mr. Daniels' pedagogy because they really were engaged in trying to achieve some level of understanding (a different kind for each girl) and needed help from Mr. Daniels, which was not forthcoming? Other learners might not have been as engaged, as concerned that they were not getting it, or as concerned about the time the conversation was taking. Many were enjoying the discussion and may have been willing to see where it took them. What kind of help could Mr. Daniels have provided, without sacrificing the benefits of discussion and engagement among the learners, yet providing some emotional and cognitive scaffolds so that the two girls did not feel so lost? In this case, may it have been more productive for Mr. Daniels not to feign ignorance but to explain why he wanted others to contribute? It may have been useful for him to indicate that there might be problems with David's and Winile's explanation and ask learners to comment specifically on these. I should point out here that there were many times during the lesson when Mr. Daniels did provide such scaffolds and used strategic combinations of a range of teacher moves (see Chap. 9), which were helpful in taking the conversation forward. So this is not a case of a teacher abdicating responsibility for teaching or not knowing how to work with reform pedagogy. Rather it illuminates the very complex nature of this kind of teaching, and how successful enactments of teaching mathematical reasoning throws up new challenges for both teacher and learner.

Making Sense of the Resistance

In this chapter, I have identified a number of different dimensions of learner resistance that were evident in one classroom, arguing that these came together to produce a complex dynamic that is not easy for the teacher to manage. Two different learners contributed differently to the openly expressed episodes of resistance, even as they built on each other's contributions to express their resistance. The teacher also contributed, by successfully pressing for the learners' own reasoning and explanations and by insisting that he would not explain for the learners. However, the learners' resistance increased when he continued to feign ignorance and as resistance mounted he reverted to expressions of traditional authority, which fuelled the learners' resistance even further.

So what exactly were the two learners resisting, and why were they resisting it? The fact that they engaged in the classroom conversation and were willing to offer their own thinking for consideration and consider the ideas of others suggests that these were not "defensive" learners in McNeill's sense; they were not opting out of the process of learning in the real sense, engaging with others' ideas. However, they both reached a moment where they did want to opt out of the current processes operating in the classroom. The fact that they did this vocally rather than retreating into silence, suggests that they still believed in the process; however, they had reached a point where the conversation was not working for them. They were not resisting all of Mr. Daniels' pedagogy, but aspects of it that did not make sense to them. Melanie wanted some authorization from the teacher because she felt she was unable to make the appropriate mathematical judgements on her own. Michelle wanted some closure, an answer to her question that she could work out and understand; but could not build on her classmates' contributions to achieve this understanding. Mr. Daniels had provided a number of resources during the lesson to help learners develop their reasoning and their understandings, resources that had helped David and Winile and other learners, but had not helped Melanie, Michelle, and possibly other learners as well who remained silent.

The literature on creating norms of participation for reform pedagogy suggests that when learners have difficulties in participating it is because they do not understand what is expected of them as they struggle to shift from traditional to new pedagogies. My analysis in this chapter suggests that this is only part of the story. Different learners will have different reasons for resisting, in this case, related to their orientations to learning. It is significant that a learner who was engaged in the mathematical practices of asking questions, relating different representations to each other, and attempting to make sense of mathematical ideas, began to resist the conversation, precisely because she was not making progress on making sense, and found some of the discussion difficult to follow. How to enable all learners to follow each other's, sometimes muddled, articulations, without constraining their thinking and participation, is a demanding task for teachers, but this chapter suggests that it is one that is worth working on. How to provide some authority, while not shutting down learners' thinking also seems to be a useful front on which to

... The idea of probing learners' ideas rather than explaining clearly might seem dishonest and a shirking of the teacher's responsibility. Additional "norm work" might be required to help learners to see this as part of the teacher's responsibility and a way of challenging learners to think more deeply.

The main argument that I have made in this chapter is that such resistance, while uncomfortable for teachers, is a normal response to reform pedagogies, and is to be preferred over the silent, safe, defensive resistance discussed at the beginning of the chapter. The fact that this resistance suggests "non-defensiveness" on the part of learners is something to celebrate. Although it may be possible to manage such instances more productively so that they do not disrupt the ongoing conversation, as a teacher successfully manages to enact reforms, so we can expect resistance. Understanding different forms of resistance, as I have attempted to do in this chapter is, probably the best way we have to develop ways of managing them. As discussed in the rest of this book, teaching for mathematical reasoning requires much more of teachers than going beyond "not telling" and establishing appropriate norms. It requires thoughtful combinations of teaching moves that respond appropriately to learner contributions. It requires managing dilemmas of teaching and it also requires some understanding of where resistance might arise, and the range of reasons for this resistance, thus adding to the difficulties and demands of such teaching and explaining its rarity.

Chapter 12
Conclusions and Ways Forward: The "Messy" Middle Ground

Much of the research on teacher change tells us that teachers either remain resistant to change (Lavi and Shriki 2008), or they embrace the rhetoric of change but their practices remain constant (Chisholm et al. 2000; Taylor 1999). Teachers are said to maintain algorithmic and procedural approaches to mathematics through lower-level tasks and strongly constrained classroom interaction; and learners very seldom engage in genuine mathematical thinking. I agree with Nolan (2008) that characterizing teaching in this way says more about researchers than teachers. For me it is more important to find ways to talk about what teachers have managed, rather than what they have not, even if the successes are small. To do this is a difficult task for researchers, but no more difficult than the task we ask of teachers when we suggest that they shift their practices.

Vygotsky's sociocultural perspective suggests the concept of the zone of proximal development as a tool to understand learning and development (Vygotsky 1978). I suggest here that it can also be used to understand teacher learning and development. Learning and development progress as new goals are set in advance of, but within the constraints of, the learner's current position. The learner can only proceed towards new goals from her current competence. Reaching ultimate learning goals requires that intermediate goals are constantly set and shifted by the learner and by those who mediate his/her journey (Wertsch 1984). So, change is a constant interaction between learners' and teachers' current and future positions, and the goals and direction of change shift as we make the journey.

In the context of curriculum reforms, the new practices that the curriculum sets out constitute a set of teaching goals. These new practices are constrained by teachers' histories, current positions, and current knowledge. Teacher change requires continuity as well as transformation; some things must stay the same in order for others to change (Slonimsky and Brodie 2006). So teacher change will be a slow, uneven, and messy process and the process will be different for different teachers. This book has described how a group of teachers, all with strong mathematical and pedagogical content knowledge, who worked together to plan tasks for mathematical reasoning, supported different kinds of interactions and engagement in their classrooms. There was variation within each teacher's practice as well as across their practices. This book has illuminated the common and different achievements of the teachers as well as the challenges that they faced as they worked to teach

K. Brodie, *Teaching Mathematical Reasoning in Secondary School Classrooms,*
DOI 10.1007/978-0-387-09742-8_12, © Springer Science+Business Media, LLC 2010

mathematical reasoning. Part of the messy middle ground of reforming practice is that achievements and challenges are often not separate from each other; achievements gives rise to new challenges. As in all areas of human endeavour, strengths can produce weaknesses and weakness can produce strengths.

To improve practice, teachers need to recognize the shifts that they have made and those they still need to make. Researchers need to find ways to describe both small and large shifts in teachers' practices and the textures and complexities of the processes of change. This book has combined perspectives of teachers and a researcher and has found ways to talk about the successes and challenges of changing practice, on two levels: Individual case studies and an overview across the cases. Each of these levels gives a different view of the teaching enterprise and how it develops. Here I draw together the findings in the book and show how they are enriched by the different views that we have taken.

Tasks and Mathematical Reasoning

One of the key shifts that reform mathematics curricula ask of teachers is to work with tasks that support mathematical reasoning. Teachers may struggle to do this for a number of reasons. They often do not have access to texts with higher-level tasks (Taylor and Vinjevold 1999) and if they do, they have to recognize the mathematics that such tasks can support, which often requires a shift in teachers' own mathematical thinking (Heaton 2000). The teachers represented in this book, working together, developed tasks that could support mathematical reasoning at different levels. They drew on a number of textbooks and materials and adapted these thoughtfully and carefully for their own classrooms.

Each set of tasks that the teachers developed reflects a range of levels of cognitive demand (Stein et al. 1996) (see Chap. 2). It is important to emphasise this range in discussions about changing curricula. It is not possible, nor desirable, for all the tasks that learners encounter to be at the higher levels. There is a place for lower-level tasks in any curriculum and these can be a point of continuity with previous practice. It is, however, important that lower-level tasks are articulated with higher-level tasks in order to develop mathematical reasoning. We need a full range of tasks in the curriculum, brought together by the teacher and learners to create opportunities for reasoning. The teachers in this book took this challenge seriously and developed tasks at a range of levels, appropriate for their classrooms.

An important issue in choosing and developing tasks is how to tailor them for a particular group of learners. Knowing that learners can only work with their current knowledge, teachers whose learners have weaker knowledge often think that higher-level tasks may be too challenging for them. Such concerns may lead to learners with weaker knowledge being denied opportunities to engage with tasks that support mathematical reasoning. In this book, the teachers were sensitive to the interactions between current and new knowledge among their particular learners. Even though the two Grade 10 teachers worked together to develop and plan tasks, they finally chose different tasks for their classrooms (with one in common), on the basis that their learners had very different levels of mathematical knowledge (see Chap. 2). Significantly, Mr. Peters chose tasks of higher-level cognitive demand that he believed his learners could tackle, even with their weaker knowledge. Mr. Peters knew his learners well and was sensitive to their weaker knowledge; however, he was not willing to compromise on teaching them mathematical reasoning and did not lower the level of the tasks and his mathematical goals as he tailored them for his learners.

Supporting Learner Contributions

Once teachers choose higher-level tasks, maintaining the level of the tasks in classroom interaction is an additional challenge, one that is not often achieved (Modau and Brodie 2008; Stein et al. 1996; Stein et al. 2000). The challenge involves supporting learner contributions and engagement with the tasks at the intended levels. The analysis in this book shows that all the teachers supported a range of learner contributions, although the distributions of the different kinds of learner contributions were different across the different classroom contexts (see Chap. 8).

An important element of reform practices is how teachers work to extend and explore learners' correct contributions. Teachers usually value correct contributions, but how they are valued distinguishes reform from traditional practice. In traditional teaching, correct answers end the conversation; in reform teaching, they are not the ultimate goal and are taken further. This book shows that all the teachers worked with correct contributions in both reform and traditional ways at different times. In some cases, they accepted and positively evaluated correct responses, ending the discussion and in others they pressed learners to justify and explain the thinking underlying their correct responses, taking the mathematics forward. The analysis suggests that working in reform ways with correct contributions can be an important point of leverage for teacher-educators to support teachers to begin shifting their practices.

A second important element of reform pedagogy is accepting, valuing, and engaging learners' incorrect and partial contributions as points where new learning can occur. Most theories of learning argue that current knowledge both constrains and enables further learning.[1] However, what this means for teachers trying to take their learners' thinking seriously is not often explored, particularly when engaging current knowledge produces errors and misconceptions. This book shows that all the teachers managed to work with learners' current knowledge to some extent, and how they managed it can be partially accounted for by their different classroom contexts. A clear finding that emerged from both the case studies and the overview

---

[1] Different theories of learning conceptualize the relationship between current and new knowledge differently (see Chap. 1).

analysis is that the teachers whose learners had weaker knowledge, experienced greater challenges in supporting learner contributions. The case studies in Chaps. 3 and 7 showed that Mr. Peters' and Mr. Nkomo's learners, who had the weakest knowledge, struggled most to engage the higher-level tasks and to justify their thinking. The overview analysis in Chap. 8 supports this and shows that Mr. Peters' learners produced more errors than learners in any of the other classrooms, including many more basic errors, while Ms. King's and Mr. Mogale's learners, who had the strongest knowledge, produced the least. Working with learner errors emerged as a key aspect of this research.

Working with Learner Errors

There are strong research traditions in the area of learners' misconceptions and in the area of teachers' take-up of reforms (see Chap. 1). However, there is little work that shows how learner errors and teachers' reform practices support each other. The analysis in this book shows that as teachers ask their learners to respond to higher-level tasks and promote discussions around these, more errors may be produced, particularly those that are task-related (appropriate errors), and more errors will become public, particularly if students feel confident to express errors that they might not have otherwise.

All learners make errors, and this is a normal part of the learning process. However, when more learners in a class have stronger knowledge, teachers can usually rely on some learners to engage with the errors of others and can expect fruitful discussion about errors. This happened in Mr. Mogale's, Mr. Daniels', and Ms. King's classrooms. However, where many learners have weaker knowledge, as learners attempt to respond to each other's errors they may produce even more errors. In Mr. Peters' classroom the errors came thick and fast, particularly basic errors. Mr. Peters successfully employed a range of teaching strategies to help learners express their thinking and engage each other's ideas. He also succeeded in supporting a genuine conversation among his weaker learners, which produced partial insights and beyond task contributions. However, his efforts were not always successful in helping learners to understand the nature of their errors and misconceptions and often brought more errors into the public arena. Since errors are immediately visible and often worrying to teachers, learning how to work with them as possibilities for new learning is an important leverage point for developing practice.

Classroom Conversations

Supporting genuine classroom conversations is another important element of reform curricula (Boaler 2002; Lampert 2001; Nystrand and Gamoran 1991). This research illuminates three important points about classroom conversations. First, it is possible to create genuine conversations in classrooms, including classrooms where learners have very weak knowledge. Second, it is not possible for all teaching to occur through genuine mathematical conversations; these are rare occurrences. Third, the teacher is always present in a conversation, even if not immediately physically present.

This book describes three examples of genuine mathematical conversation, one in each of Mr. Mogale's, Mr. Daniels', and Mr. Peters' classrooms. In Mr. Mogale's and Mr. Daniels' classrooms the learners had stronger knowledge and the conversations both supported and were supported by many of the partial insights that were evident in those classrooms. Mr. Peters and his learners also managed to generate and sustain a genuine conversation, about whether zero is a positive or negative number. The analysis shows that a number of learners in this class reasoned usefully and appropriately as they tried to resolve this issue and that Mr. Peters' pedagogy was central in supporting this engagement.

The analysis in this book also shows that it is not always possible to have genuine mathematical conversation in classrooms. These happen on rare occasions and require particular conditions of possibility. They take hard work and teacher insight to instigate, maintain, and end appropriately (Brodie 2007b). When they do happen, they are exciting, particularly because learners build on each others' reasoning and the collaborative and dialogic (Mortimer and Scott 2003) nature of learning becomes visible and is supported. So, this work argues that while teachers can and should recognize possible occasions for conversations and exploit these, not all of teaching and learning can consist of conversations.

This work also takes issue with a prevalent idea in South Africa and elsewhere that genuine conversations in classrooms can only occur without the teacher being present. This idea is supported by a misinterpreted maxim of constructivism: "Learners learn on their own". In bringing constructivism and sociocultural theories together, this maxim can be rephrased more usefully as "only the learner can learn and s/he cannot learn alone". Drawing further on Vygotsky's work and Mr. Mogale's case study, we have shown that fruitful discussion can only occur without the teacher's immediate presence if the teacher has mediated such discussions previously and provided the learners with the conversational and conceptual tools to interact with each other. Mr. Mogale shows how his learners internalized his practices, which in turn supported their interactions both with him and with each other.

Maintaining the IRE/F

Arguments for teaching through conversations are usually supported by arguments for reducing the predominance of the IRE/F exchange structure in classrooms. The work in this book presents a different argument. In all the classrooms in this study, the IRE/F structure remained predominant, in form and often in function as well. I, therefore, argue that it is more useful to open up possibilities for working within the IRE/F than suggesting that teachers avoid it entirely (see also Wells 1999). Teachers do need to evaluate and/or give feedback on learners' contributions and are well practiced at doing so. The analysis in Chap. 9 shows that teachers can and do respond in a range of ways within the IRE/F format and that opening up a wider range of possibilities for teachers can be generative in supporting them to develop new practices. Within the IRE/F, the teachers pressed learners' thinking and facilitated and maintained discussion, and even when they inserted their own mathematical ideas they helped learners to move their thinking forward (see also Lobato et al. 2005). Perhaps the most problematic move in the teachers' repertoire was the "elicit" move, because it tended to narrow learners' thinking. However, even with this move, there are examples where such eliciting functioned to generate more and better discussion.

Supporting all Learners to Participate

This book has shown that supporting some learners to express their mathematical thinking is a challenging task for teachers. For teachers who are successful in doing this, encouraging all learners to participate, particularly in classrooms of up to 45 learners, is an ongoing challenge. While all of the teachers in this study found ways to increase their learners' participation in the lessons and more learners participated more often across all of the classrooms, there were some learners in each class who did not participate at all during the week. While some learners may prefer to engage by listening, it is likely that some were not engaged at all. This work illuminated the challenges that all of the teachers experienced as they worked to increase participation.

Mr. Nkomo's case study illuminates two important points: First, just how difficult it is to support learners to participate at higher levels of mathematical thinking; and second that teachers can learn to do this. Mr. Nkomo's reflections on his practice show how his first interactions with learners did not enable them to interact at all with each other or with him. Later, he managed some interaction but not at higher levels of thinking. Finally, he managed to get some learners to think and talk about each other's ideas in ways that helped to develop the ideas. So it takes time and support for teachers to interact with their learners in useful ways.

Even the teachers who were more successful in generating conversations experienced challenges in encouraging all learners to participate. Precisely because of their success, they experienced dilemmas in relation to learner participation. In Chap. 10, I showed how Mr. Peters and Mr. Daniels sometimes struggled with which contributions to take up for further discussion and which to ignore. When many learners contributed, not all could be heard all of the time. Mr. Peters experienced an additional dilemma about whether or not to press a shy learner to contribute. He persevered and used his authority to assign competence to the learner, encouraging him to participate. Teachers are often reluctant to embarrass learners by calling them to participate. This study has shown that teachers can work sensitively with learners' feelings while still supporting them to participate.

Learner Resistance

There is very little work on learner resistance to changing teaching. The reform argument is that reform practices will reduce learner resistance to schooling. However, it is possible, as shown in Chap. 11 that shifting towards reform practices can produce new forms of resistance. Different learners can resist for different reasons, and the teacher may unwittingly contribute to learner resistance as s/he struggles to deal with it. Although explicit learner resistance is distressing and extremely challenging for teachers, my analysis suggests that such resistance can be seen as a sign of success and can be understood and managed appropriately by teachers. As teacher educators, it is important to alert teachers to the possibilities of such resistance, and to think about ways of working with it.

Conclusions

The teachers' achievements and challenges illuminated in this study constitute the messy middle ground of reforming practice. All these teachers developed broader understandings of mathematical knowledge and reasoning, they all chose tasks to support mathematical reasoning, they all to some extent managed to shift interaction patterns in their classrooms to accommodate learner contributions and ideas, and they all found ways to think about these and bring them into their lessons. The challenges emerged in the on-the-spot interactions with particular learners' thinking, moving between old and new practices, and maintaining the levels of mathematical reasoning among all learners. This book has identified a number of leverage points for changing practice and presented a language of description to enable teachers, teacher educators, and researchers to talk about these. It is only in describing and analyzing the challenges and unevenness of change that we will make progress towards better mathematics learning for all learners.

One of the ongoing themes of this book has been how the five teachers worked together and with me to form a community to support changing practices. This community provided a safe space to try out new ideas, to take risks and to reflect on these in ongoing and systematic ways. The teacher's research projects, conducted as collaborative action research projects provided the space for exploration and systematic reflection. Some of the teachers had been working with these ideas before we began this project and all have continued to do so. Talking and working together as a small community has made it possible to initiate a broader conversation about these important issues through this book.

The key aim of this book is to capture the textures and complexities of teaching practices as teachers work to adapt their practice to new curriculum ideas, in this case, the teaching of mathematical reasoning. In trying to find ways to talk about practice we have come to understand more deeply the challenges involved in shifting towards reform-oriented teaching. We have come to realize that meaningful change in teaching takes time, imagination, courage, and honest reflection on what works and what does not, and that these qualities are much easier to talk about than to achieve. We have also confirmed our assumption that teaching is a continuous quest for improved knowledge and improved practice. Educational reformers and teachers often create lofty goals, like the goals of reform mathematics curricula. We do this because we are inspired to improve education for subsequent generations and because in striving for more challenging goals we are in a better position to achieve more reasonable ones. We need to recognize that important goals take a long time to achieve – which is why they have not yet been achieved. In questing for "perfection," we must realize that we may never reach it and find ways to talk about and share what we have indeed achieved, so that our quest can continue.

Appendix

Grade 11   **Quadratic functions in a different form**

Name: _____

<u>Activity 1</u>   **Moving Left and Right**

The graph and table of the function $y = x^2$ is shown below.

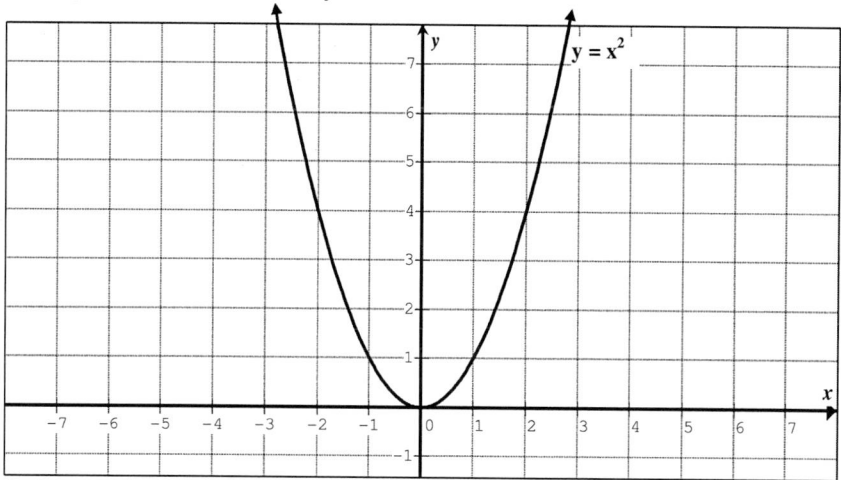

x-values of $y = x^2$	-3	-2	-1	0	1	2	3
y-values of $y = x^2$	9	4	1	0	1	4	9

1. Copy the graph of $y = x^2$ onto a sheet of firm plastic.

  1.1. a) Move the copy of the graph $y = x^2$ three units to the right.

  b) Each point on the graph $y = x^2$ has moved to a new position. Compare the turning point of $y = x^2$ to the turning point of the new graph. Choose other points on the graph of $y = x^2$ and compare them to corresponding points on the new graph. What do you observe about the corresponding points of the two graphs? Record any changes you observe in the rows provided in the table above.

  1.2. a) Move the copy of the graph $y = x^2$ four units to the left.

  b) Each point on the graph $y = x^2$ has moved to a new position. Compare the turning point of $y = x^2$ to the turning point of the new graph. Choose other points on the graph of $y = x^2$ and compare them to corresponding points on the new graph. What do you observe about the corresponding points of the two graphs? Record any changes you observe in the rows provided in the table above.

Activity 2   **Moving Right and Left Again**

Name: _____

The graph and table of the function $y = x^2$ is shown below. The graph shifted 3 units to the right and the graph shifted 4 units to the left are also shown on the same set of axes. Study the graphs below and then answer the questions that follow.

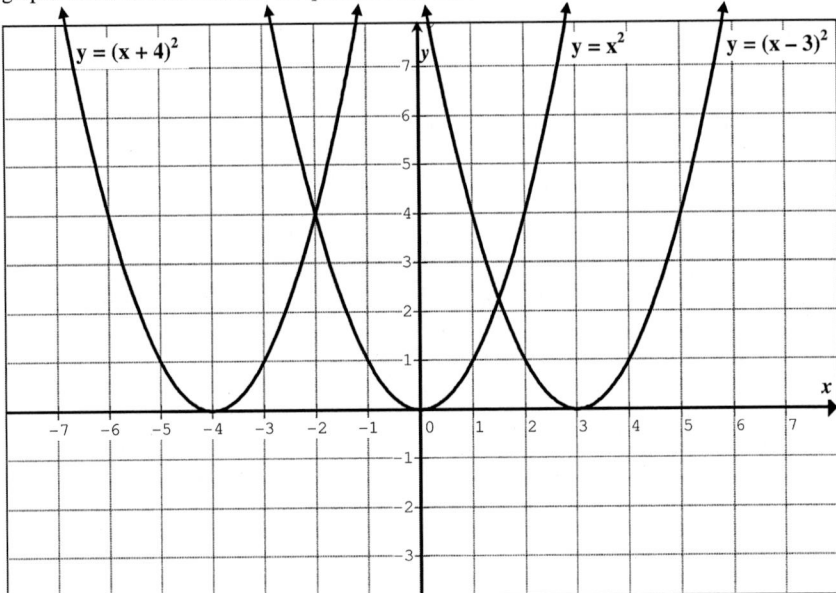

1. Complete the table

Function	Minimum Value	Symmetry Axis	Turning Point
$y = x^2$			
$y = (x + 4)^2$			
$y = (x - 3)^2$			

2. Discuss with a partner how these graphs differ and how they are the same.
3. Compare the x and y-values of $y = x^2$ to the x and y-values of $y = (x + 4)^2$ and $y = (x - 3)^2$. What do you observe?
4. How can we obtain the graph of $y = (x + 4)^2$ from the graph of $y = x^2$?
5. How can we obtain the graph of $y = (x - 3)^2$ from the graph of $y = x^2$?
6. The graphs are all of the form $y = a(x - p)^2$. How does the value of p affect the graph?
7. Provide two of your own examples by:
    a) completing the last two rows in the above table, and
    b) sketching the two graphs on the same set of axes provided

__Activity 3__   **Moving Up and Down**

Name: _____

Study the graphs below and answer the questions that follow

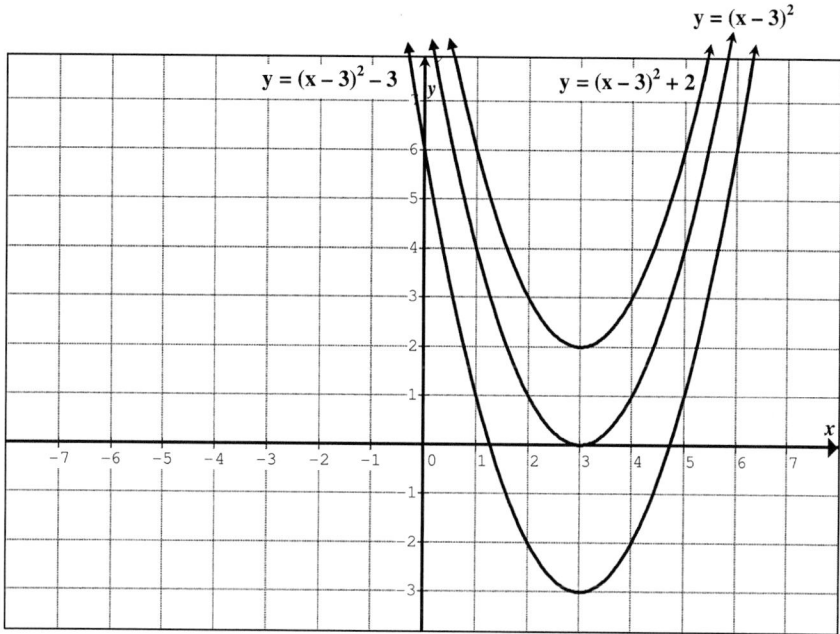

1. Complete the table

Function	Minimum Value	Symmetry Axis	Turning Point
$y = (x - 3)^2$			
$y = (x - 3)^2 + 2$			
$y = (x - 3)^2 - 3$			

2. Discuss with a partner how these graphs differ and how they are the same.
3. Compare the x and y-values of $y = (x - 3)^2$ to the x and y-values of $y = (x - 3)^2 + 2$ and $y = (x - 3)^2 - 3$. What do you observe?
4. How can we obtain the graph of $y = (x - 3)^2 + 2$ from the graph of $y = (x - 3)^2$?
5. How can we obtain the graph of $y = (x - 3)^2 - 3$ from the graph of $y = (x - 3)^2$?
6. The graphs are all of the form $y = a(x - p)^2 + q$. How does the value of q affect the graph?
7. Provide two of your own examples by:
   c) completing the last two rows in the above table, and
   d) sketching the two graphs on the same set of axes provided

Grade 10 Tasks: Ms. King

Task 1

Work in groups but hand in your own work. Work in the space provided below.
Consider the following conjecture: x^2+1 can never be 0.
(a) Use a logical argument to convince someone else why the conjecture is either true or false for any real value of x.
(b) What is the smallest value of x^2+1? Explain how you know.

Task 2 (work on this sheet)

(a) In the following list, the numbers on the right are related to those on the left:

$$
\begin{array}{ccc}
x & & y \\
1 & \rightarrow & 1 \\
2 & \rightarrow & 4 \\
3 & \rightarrow & 9 \\
4 & \rightarrow & 16
\end{array}
$$

(b) Can you find the rule that relates these numbers? Describe this rule in words.
(c) Can you write this rule mathematically?
(d) Ask your teacher to show you some other ways of writing this rule in mathematical notation. Then describe in words what each of the different notations means.

Task 3

In order to talk about the above rule, we need to give it a name. We sometimes call it f, and we write it mathematically as $f(x)=x^2$. This means that when we apply the rule f to the number x, we get x^2. When we write $f(2)$, we mean "apply the rule f to the number 2."
So $f(2)=2^2=4$.
Similarly, $f(3)=3^2=9$.

Work out the following:
 (i) $f(4)=$
 (ii) $f(5)=$
 (iii) $f(-2)=$
 (iv) $f(-1)=$
 (v) $f(a)=$
 (vi) $f(a+h)=$
 (vii) $f(x+1)=$

What do you think $f(a+h)$ means?

Definition of a function

We are now ready for a working definition of a function:

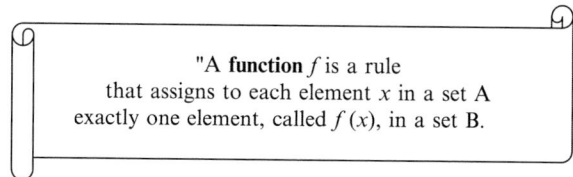

"A **function** f is a rule
that assigns to each element x in a set A
exactly one element, called $f(x)$, in a set B.

Explain, in your own words, what you understand a function to be. Draw a picture if this will help you to explain.

N.B.

We often represent functions with the letters f, g, h.

The letter in brackets after the f, g, h refers to the variable.

We could write $f(x)=2x+1$, but *would not write* $g(x)=z+3$, as the variables on the left and right hand sides do not correspond.

Task 4

If $g(x)=2x^2+3x-1$, evaluate the following:

(a) $g(1)$
(b) $g(-1)$
(c) $g(2)$
(d) $g(-2)$
(e) $g(a)$
(f) $g(-a)$
(g) $g(a+h)$
(h) $\dfrac{g(a+h)-g(a)}{h}$; $h\neq 0$

Task 5

Consider the following statement: "$f(n)=n^2-n+11$ is a prime number for all natural numbers n."

(a) List the first five natural numbers.
(b) Determine $f(1), f(2), f(3), f(4)$ etc.,
(c) Is the above statement true? i.e., Does $f(n)$ always generate a prime number?
(d) Try to justify/prove your answer in (c) above.

Grade 10 Tasks: Mr. Peters

Task 1

Consider the following conjecture: "$x^2 + 1$ can never be zero"!
  Prove whether this statement is true or false if $x \in R$?

Task 1 (B)

(a) Complete the table.

x	−3	−2	−1	0	1	2	3
x^2							
$2x^2$							
$-x^2$							
$-2x^2$							

(b) Study rows 1–4. What conclusions can be drawn from the observation?
(c) If $x = \frac{1}{2}$ or $\sqrt{7}$, would your conclusions be true?
(d) Referring back to task 1, "$x^2 + 1$ can never be zero"!
    Is this statement is true or false?
(e) Justify or explain why you say so.

Task 2

State whether the following expressions in terms of x are:

(i) Always ≥ 0
(ii) Always ≤ 0
(iii) Sometimes positive, sometimes zero, sometimes negative, depending on the
     value of x.

(a) x	(b) $-2x$	(c) x^2
(d) $3x^2$	(e) $-x^2$	(f) $(x+1)^2$
(g) $-(x+2)^2$	(h) $2(x-3)^2$	

References

Adler J, Reed Y (eds) (2002) Challenges of teacher development: an investigation of take-up in South Africa. Van Schaik, Pretoria

Alro H, Skovsmose O (2004) Dialogue and learning in mathematics education. Kluwer, Dordrecht

Association for Mathematics Education of South Africa (2000) Submission by the Mathematics Education Community to the Council of Education Ministers

Association for the Development of Education in Africa (ADEA) (2003) Pedagogical renewal and teacher development in sub-Saharan Africa: a thematic synthesis: document prepared for ADEA biennial meeting, Grand Baie, Mauritius, 3–6 December 2003

Balacheff N (1991) The benefits and limits of social interaction: the case of proof. In: Bishop A, Mellin-Olsen S, van Dormolen J (eds) Mathematical knowledge: its growth through teaching. Kluwer, Dordrecht, pp 175–192

Ball DL (1993) With an eye on the mathematical horizon: dilemmas of teaching elementary school mathematics. Elem Sch J 93(4):373–397

Ball DL (1996) Teacher learning and the mathematics reforms: what we think we know and what we need to learn. Phi Delta Kappan 77:500–508

Ball DL (1997) What do students know? Facing challenges of distance, context and desire in trying to hear children. In: Biddle BJ, Good TL, Goodson IF (eds) International handbook of teachers and teaching. Kluwer, Dordrecht, pp 769–818

Ball DL (2003) Mathematical proficiency for all students: towards a strategic research and development program in mathematics education. (DRU-2773-OERI). Rand mathematics study panel for office of educational research and improvement, Santa Monica, CA

Ball DL, Bass H (2003) Making mathematics reasonable in school. In: Kilpatrick J, Martin WG, Schifter DE (eds) A research companion to principles and standards for school mathematics. National Council of Teachers of Mathematics, Reston, VA, pp 27–44

Ball DL, Cohen DK (1999) Developing practice, developing practitioners: towards a practice-based theory of professional education. In: Darling-Hammand L, Sykes G (eds) Teaching as the learning profession. Jossey-Bass, San Francisco, pp 3–32

Barnes D, Todd F (1977) Communication and learning in small groups. Routledge and Kegan Paul, London

Bauersfeld H (1980) Hidden dimensions in the so-called reality of a mathematics classroom. Educ Stud Math 11:23–41

Bauersfeld H (1988) Hidden dimensions in the so-called reality of a mathematics classroom. Educ Stud Math 11:23–41

Bennie K (2006) Focus on mathematics grade 11. Maskew Miller Longman, Cape Town

Ben-Zeev T (1996) When erroneous mathematical thinking is just as correct: the oxymoron of rational erros. In: Sternberg RJ, Ben-Zeev T (eds) The nature of mathematical thinking. Lawrence Erlbaum Associates, Mahwah, NJ, pp 55–79

Ben-Zeev T (1998) Rational errors and the mathematical mind. Rev Gen Psychol 2(4):366–383

Biggs JB, Collis KF (1982) Evaluating the quality of learning: the SOLO taxonomy. Academic, New York

Boaler J (1997) Experiencing school mathematics: teaching styles, sex and setting. Open University Press, Buckingham

Boaler J (2002) Learning from teaching: exploring the relationship between reform curriculum and equity. J Res Math Educ 33(4):239–258

Boaler J (2004) Promoting relational equity in mathematics classrooms: Important teaching practices and their impact on student learning. Paper presented at the international congress on mathematics education, Copenhagen, July 2004

Boaler J, Brodie K (2004) The importance, nature and impact of teacher questions. In: McDougall DE, Ross JA (eds) Proceedings of the 26th annual meeting of the North American chapter of the international group for the psychology of mathematics education, vol 2. Ontario Institute of Studies in Education/University of Toronto, Toronto, pp 773–781

Boaler J, Greeno JG (2000) Identity, agency and knowing in mathematics worlds. In: Boaler J (ed) Multiple perspectives on mathematics teaching and learning. Ablex, Westport, CT, pp 171–200

Boaler J, Humphreys C (2005) Connecting mathematical ideas: middle school cases of teaching and learning. Heinemann, Portsmouth, NH

Brodie K (1999) Working with pupils' meanings: Changing practices among teachers enrolled on an in-service course in South Africa. In: Zaslavsky O (ed) Proceedings of the 23rd conference of the international group for the psychology of mathematics education, vol 2. Israel Institute of Technology, Haifa, pp 145–152

Brodie K (2000) Constraints in learner-centred teaching: a case study. J Educ 25:131–160

Brodie K (2003) Being a facilitator and a mediator in mathematics classrooms: a multidimensional task. In: Jaffer S, Burgess L (eds) Proceedings of the 9th national congress of the association for mathematics education of South Africa. University of Cape Town, Cape Town, pp 135–146

Brodie K (2004a) Re-thinking teachers' mathematical knowledge: a focus on thinking practices. Perspect Educ 22(1):65–80

Brodie K (2004b) Working with learner contributions: coding teacher responses. In: McDougall DE, Ross JA (eds) Proceedings of the 26th annual meeting of the North American chapter of the international group for the psychology of mathematics education, vol 2. Ontario Institute of Studies in Education/University of Toronto, Toronto, pp 689–697

Brodie K (2005) Textures of talking and thinking in secondary mathematics classrooms. Unpublished Phd Dissertation, Stanford University

Brodie K (2006) Teaching mathematics for equity: learner contributions and lesson structure. Afr J Res Math Sci Technol Educ 10(1):13–24

Brodie K (2007a) Dialogue in mathematics classrooms: beyond question and answer methods. Pythagoras 66:3–13

Brodie K (2007b) Teaching with conversations: beginnings and endings. Learn Math 27(1):17–23

Brodie K (2007c) The mathematical work of teaching: beyond distinctions between traditional and reform. In: Mutimucuio I, Cherinda M (eds) Proceedings of the 15th annual meeting of the Southern African association for research in mathematics, Science and Technology Education (SAARMSTE). Eduardo Mondlane University, Maputo

Brodie K (2008) Describing teacher change: Interactions between teacher moves and learner contributions. In: Matos JP, Valero P, Yakasuwa K (eds) Proceedings of the fifth international mathematics education and society conference (MES5), pp 31–50

Brodie K (in press) Pressing dilemmas: meaning making and justification in mathematics teaching. J Curriculum Stud

Brodie K, Pournara C (2005) Toward a framework for developing and researching groupwork in South African mathematics classrooms. In: Vithal R, Adler J, Keitel C (eds) Mathematics education research in South Africa: possibilities and challenges. Human Sciences Research Council, Pretoria

Brodie K, Lelliott T, Davis H (2002) Developing learner-centred practices through the FDE programme. In: Adler J, Reed Y (eds) Challenges of teacher development: an investigation of take-up in South Africa. Van Schaik, Pretoria, pp 94–117

Brodie K, Shahan E, Boaler J (2004) Teaching mathematics and social justice: multidimensionality and responsibility. Paper presented at the international congress of mathematics education (ICME10), Copenhagen, Denmark

Brodie K, Coetzee K, Modau AS, Molefe N (2005) Teachers and academics collaborating on research on mathematics classroom interaction. Paper presented at the international commission on mathematics instruction (ICMI) Africa regional conference, University of the Witwatersrand, Johannesburg, 22–25 June 2005

Brodie K, Sanni R, Jina Z, Modau AS, Molefe N (2007) Relationships between changing mathematics curricula and changing pedagogy in two African contexts. Symposium presented at the learning conference, Johannesburg, South Africa, 26–29 June 2007

Brousseau G, Gibel P (2005) Didactical handling of students' reasoning processes in problem solving situations. Educ Stud Math 59:13–58

Brown JS, Collins A, Duguid P (1989) Situated cognition and the culture of learning. Educ Res 18(1):32–42

Carraher DW (1996) Learning about fractions. In: Steffe P, Nesher P, Cobb P, Goldin G (eds) Theories of mathematical learning. Lawrence Erlbaum, New Jersey, pp 241–264

Cazden CB (1988) Classroom discourse: the language of teaching and learning. Heinemann, Portsmouth, NH

Centre for Development and Enterprise (2007) Doubling for growth: addressing the maths and science challenges in South Africa's schools. Centre for Development and Enterprise, Parktown

Chazan D (1993) High school geometry students' justification for their views of empirical evidence and mathematical proof. Educ Stud Math 24:359–387

Chazan D (2000) Beyond formulas in mathematics and teaching: dynamics of the high school algebra classroom. Teachers' College Press, New York

Chazan D, Ball DL (1999) Beyond being told not to tell. Learn Math 19(2):2–10

Chick JK (1996) Safe-talk: collusion in apartheid education. In: Hywel C (ed) Society and the language classroom. Cambridge University Press, Cambridge, pp 21–39

Chisholm L, Volmink J, Ndhlovu T, Potenza E, Mahomed H, Muller J et al (2000) A South African curriculum for the twenty first century. Report of the review committee on curriculum 2005. Department of Education, Pretoria

Christie P, Potterton M (1997) School development in South Africa: strategic interventions for quality improvement in South African schools. Joint Education Trust, Johannesburg

Chung S, Walsh DJ (2000) Unpacking child-centredness: a history of meanings. J Curriculum Stud 32:215–234

Cobb P (2000) The importance of a situated view of learning to the design of research and instruction. In: Boaler J (ed) Multiple perspectives on mathematics teaching and learning. Ablex, Westport, CT, pp 45–82

Cohen EG (1994) Designing groupwork: strategies for the heterogenous classroom. Teachers' College Press, New York

Confrey J (1990) A review of research on student conceptions in mathematics, science and programming. In: Cazden CB (ed) Review of research in education, vol 16. American Educational Research Association, Washington, pp 3–56

Confrey J, Kazak S (2006) A thirty-year reflection on constructivism in mathematics education in PME. In: Gutierrez A, Boero P (eds) Handbook of research on the psychology of mathematics education: past, present and future. Sense Publishers, Rotterdam

Copes L (1982) The Perry development scheme: a metaphor for learning and teaching mathematics. Learn Math 3(1):38–44

Crook C (1994) Human cognition as socially grounded computers and the collaborative experience of learning. Routledge, London, pp 30–51

Cuban L (1993) How teachers taught: constancy and change in American classrooms. Teachers' College Press, New York

Daniels H (2001) Vygotsky and pedagogy. Routledge/Falmer Press, London

Davis RB (1988) The world according to McNeill. J Math Behav 7:51–78

Davis B (1997) Listening for differences: an evolving conception of mathematics teaching. J Res Math Educ 28(3):355–376

Davis PJ, Hersh R (1981) The mathematical experience. Houghton Mifflin, Boston

Davis RB, Maher CA (1997) How students think, the role of representation. In: English LD (ed) Mathematical reasoning: analogies, metaphorsm images. Lawrence Erlbaum Associates, Hillsdale, NJ

De Villiers M (1990) The role and function of proof in mathematics. Pythagoras 24:17–24

Department of Education (1997) Curriculum 2005: lifelong learning for the 21st century. National Department of Education, Pretoria

Department of Education (2000) Curriculum 2005: towards a theoretical framework. Department of Education, Pretoria

Department of Education (2001) National strategy for mathematics, science and technology in general and further education and training. National Department of Education, Pretoria

Department of Education (2003) National curriculum statement for mathematics (Grades 10–12). National Department of Education, South Africa, Pretoria

Department of Education (2007) National assessment report. National Education Infrastructure Management System (NEMIS), Pretoria

Department of Education (2008) Education statistics in South Africa: 2006. Department of Education, Pretoria

Douek N (2002) Context complexity and argumentation. In: Cockburn AD, Nardi E (eds) Proceedings of the 26th annual meeting of the international group of the psychology of mathematics education (PME), vol 2. University of East Anglia, Norwich, pp 297–305

Douek N (2005) Communication in the mathematics classroom: argumentation and development of mathematical knowledge. In: Chronaki A, Christiansen IM (eds) Challenging perspectives on mathematics classroom communication. Information Aga Publishing, Greenwich, CT, pp 145–172

Edwards D, Mercer N (1987) Common knowledge: the growth of understanding in the classroom. Routledge, London

Erlwanger SH (1975) Case studies of children's conceptions of mathematics part 1. J Child Math Behav 1(3):157–283

Ernest P (1991) The philosophy of mathematics education. Falmer Press, London

Fischbein E (1987) Intuition in science and mathematics. Kluwer, Dordrecht

Fleisch B (2007) Primary education in crisis: why South African schoolchildren underachieve in reading and mathematics. Juta, Cape Town

Fleisch B, Shindler J, Perry H (2008) Who is out of school? Evidence from the community survey 2007, South Africa paper presented at the Kenton education association, Broederstroom, 26–29 October 2008

Forman EA, Ansell E (2002) Orchestrating the multiple voices and inscriptions of a mathematics classroom. J Learn Sci 11:251–274

Fraivillig JL, Murphy LA, Fuson KC (1999) Advancing children's mathematical thinking in everyday mathematics classrooms. J Res Math Educ 2:148–170

Garuti R, Boero P (2002) Interiorisation of *forms* of argumentation. In: Cockburn AD, Nardi E (eds) Proceedings of the 26th annual meeting of the international group of the psychology of mathematics education (PME), vol 2. University of East Anglia, Norwich, pp 408–415

Glachan M, Light P (1982) Peer-interaction and learning: can two wrongs make a right? In: Butterworth G, Light P (eds) Social cognition: studies of the development of understanding. Harvester Press, Brighton, pp 238–262

Greeno JG, MMAP (1998) The situativity of knowing, learning and research. Am Psychol 53:5–26

Greeno JG, Collins AM, Resnick LB (1996) Cognition and learning. In: Berliner DC, Calfee RC (eds) Handbook of educational psychology. Simon and Shuster, MacMillan, New York, pp 15–46

Hanna G, Jahnke N (1996) Proof and proving. In: Bishop AJ, Clements K, Keitel C, Kilpatrick J, Laborde C (eds) International handbook of mathematics education. Kluwer, Dordrecht

Hatano G (1996) A conception of knowledge acquisition and its implications for mathematics education. In: Steffe P, Nesher P, Cobb P, Goldin G, Greer B (eds) Theories of mathematical learning. Lawrence Erlbaum, New Jersey, pp 197–217

Hayes D, Mills M, Christie P, Lingard B (2006) Teachers and schooling making a difference: productive pedagogies, assessment and performance. Allen & Unwin, Crows Nest, NSW

Heaton R (2000) Teaching mathematics to the new standards: relearning the dance. Teachers' College Press, New York

Hedegaard M (1990) The zone of proximal development as basis for instruction. In: Moll L (ed) Vygotsky and education. Cambridge University Press, Cambridge, pp 171–195

Herrenkohl LR, Wertsch JV (1999) The use of cultural tools: mastery and appropriation. In: Siegel I (ed) Development of mental representations. Heineman, Mahwah, NJ, pp 416–435

Hiebert J, Lefevre P (1986) Conceptual and procedural knowledge in mathematics: an introductory analysis. In: Hiebert J (ed) Conceptual and procedural knowledge: the case of mathematics. Lawrence Erlbaum, Hillsdale, NJ, pp 1–27

Howie SJ, Hughes CA (1998) Mathematics and science literacy of final-year school students in South Africa: TIMSS South Africa. Human Sciences Research Council, Pretoria

Hufferd-Ackles K, Fuson KC, Sherin MG (2004) Describing levels and components of a math-talk learning community. J Res Math Educ 35(2):81–116

Jaffer S, Johnson Y (2004) Maths for all learner's activity book: grade 10. Macmillan Boleswa, Cape Town

Jansen J (1999) A very noisy OBE: the implementation of OBE in grade 1 classrooms. In: Jansen J, Christie P (eds) Changing curriculum: studies on outcomes-based education in South Africa. Juta, Cape Town, pp 203–217

Jaworski B (1994) Investigating mathematics teaching: a constructivist enquiry. Falmer Press, London

Jina Z, Brodie K (2008) Teacher questions and interaction patterns in the new and old curriculum: a case study. In: Polaki MV, Mokuku T, Nyabanyaba T (eds) Proceedings of the sixteenth annual congress of the Southern African association for research in mathematics, science and technology education (SAARMSTE). Maseru, Lesotho

Johnson Y, McBride S, MacKay R, Brundrit S, Bowie L (2006) Maths for all learner's acxtivity book: grade 11. MacMillan, Cape Town

Kazemi E, Stipek D (2001) Promoting conceptual thinking in four upper-elementary mathematics classrooms. Elem Sch J 102:59–80

Keitel C (2000) Cultural diversity, internationalisation and globalisation: challenges or perils for mathematics education. In: Mahlomaholo S (ed) Proceedings of the eighth annual conference of the Southern African association for research in mathematics and science education (SAARMSE), pp 21–36

Kilpatrick J, Swafford J, Findell B (eds) (2001) Adding it up: helping children learn mathematics. National Academy Press, Washington, DC

Kitchen RS, DePree J, Celedon-Pattichis S, Brinkerhoff J (2007) Mathematics education at highly effective schools that serve the poor: strategies for change. Lawrence Erlbaum Associates, Mahwah, NJ

Kline M (1980) The loss of certainty. Oxford University Press, Oxford

Kloss RJ (1994) A nudge is best: helping students through the Perry scheme of intellectual development. Coll Teach 42(4):151–158

Krummheuer G (1995) The ethnography of argumentation. In: Cobb P, Bauersfeld H (eds) The emergence of mathematical meaning. Lawrence Erlbaum, Hillsdale, NJ

Lakatos I (1976) Proofs and refutations. Cambridge University Press, Cambridge

Lampert M (1985) How do teachers manage to teach? Harv Educ Rev 55(2):178–194

Lampert M (2001) Teaching problems and the problems of teaching. Yale University Press, New Haven

Lave J (1993) Situating learning in communities of practice. In: Resnick LB, Levine JM, Teasley SD (eds) Perspectives on socially shared cognition. American Psychological Association, Washington, DC, pp 63–85

Lave J (1996) Teaching, as learning, in practice. Mind Cult Act 3(3):149–164

Lave J, Wenger E (1991) Situated learning: legitimate peripheral participation. Cambridge University Press, Cambridge

Lave J, Wenger E (1999) Legitimate peripheral participation in communities of practice. In: McCormick R, Paechter C (eds) Learning and knowledge. Paul Chapman and the Open University, London

Lavi I, Shriki A (2008) Social and didactical aspects of engagement in innovative learning and teaching methods: the case of Ruth. In: Matos JP, Valero P, Yakasuwa K (eds) Proceedings of the fifth international mathematics education and society conference (MES5), Lisbon, pp 330–339

Lerman S (1998) Learning as social practice: an appreciative critique. In: Watson A (ed) Situated cognition and the learning of mathematics. Centre for mathematics education research, University of Oxford, Oxford, pp 33–42

Lobato J, Clarke D, Ellis AB (2005) Initiating and eliciting in teaching: a reformulation of telling. J Res Math Educ 36(2):101–136

Lubienski ST (2000) Problem solving as a means towards mathematics for all: an exploratory look through a class lens. J Res Math Educ 31(4):454–482

Mason J, Burton L, Stacey K (1982) Thinking mathematically. Addison-Wesley, London

McClain K, Cobb P (2001) An analysis of development of sociomathematical norms in one first-gade classroom. J Res Math Educ 32(3):236–266

McKinney C (2004) A little hard piece of grass in your shoe: understadning student resistance to critical literacy in post-apartheid South Africa. South Afr Linguist Appl Lang Stud 22(1&2):63–73

McKinney C (2005) A balancing act: ethical dilemmas of democratic teaching within critical pedagogy. Educ Action Res 13(3):375–391

McLachlan ID, Ryan DJ (1994) A.I.M.S. in the classroom. Math Teach 87:364–370

McNeil LM (1981) Negotiating classroom knowledge: beyond achievement and scoialisation. J Curriculum Stud 13(4):313–328

McNeil LM (1999) Contradictions of control: school structure and school knowledge. Routledge, New York

Mehan H (1979) Learning lessons: social organisation in the classroom. Harvard University Press, Cambridge, MA

Mercer N (1995) The guided construction of knowledge: talk between teachers and learners. Multilingual matters, Clevedon

Modau AS, Brodie K (2008) Understanding a teacher's choice of mathematical tasks in the old and new curriculum. In: Polaki MV, Mokuku T, Nyabanyaba T (eds) Proceedings of the sixteenth annual congress of the Southern African association for research in mathematics, science and technology education (SAARMSTE). Maseru, Lesotho

Moll I (2000) Clarifying constructivism. Paper presented at the Kenton educational association conference, Port Elizabeth, October 2000

Mortimer FE, Scott PH (2003) Meaning making in secondary school classrooms. Open University Press, Maidenhead

Moses R, Cobb CE Jr (2001) Radical equations: math literacy and civil rights. Beacon Press, Boston

Motala S, Dieltiens V (2008) Education access to schooling in South Africa – a district perspective. Paper presented at the Kenton education association, Broederstroom, 26–29 October 2008

Motala S, Perry H (2002) The 2001 senior certificate examination. Q Rev Educ Train South Afr 9(1):2–10

Nathan MJ, Knuth EJ (2003) A study of whole classroom mathematical discourse and teacher change. Cogn Instr 21(2):175–207

Nesher P (1987) Towards and instructional theory: the role of students' misconceptions. Learn Math 7(3):33–39

Nolan K (2008) Theory-Practice transitions and dis/positions in secondary mathematics teacher education. In: Matos JP, Valero P, Yakasuwa K (eds) Proceedings of the fifth international mathematics education and society conference (MES5), pp 406–415

Nystrand M, Gamoran A (1991) Student engagement: when recitation becomes conversation. In: Waxman HC, Walberg HJ (eds) Effective teaching: current research. McCutchan Publishing Corporation, Berkeley, CA, pp 257–276

Nystrand M, Gamoran A, Kachur R, Prendergast C (1997) Opening dialogue. Teachers College Press, New York

O'Connor MC, Michaels S (1996) Shifting participant frameworks: orchestrating thinking practices in group discussion. In: Hicks D (ed) Discourse, learning and schooling. Cambridge University Press, New York, pp 63–103

O'Sullivan M (2004) The reconceptualisation of learner-centred approaches: a Namibian case study. Int J Educ Dev 24:585–602

Osborne MD (1997) Balancing individual and the group: a dilemma for the constructivist teacher. J Curriculum Stud 29(2):183–196

Piaget J (1964) Development and learning. In: Ripple RE, Rockcastle VN (eds) Piaget rediscovered. Cornell University Press, Ithaca

Piaget J (1968) Six psychological studies. Vintage Books, New York

Piaget J (1975) The development of thought: equilibration of cognitive structures. Basil Blackwell, Oxford

Polya G (1994/1990) How to solve it. Penguin, USA

Reddy V (2006) Mathematics and science achievement at South African schools in 2003. Human Sciences Research Council Press, Cape Town

Rowell JA (1989) Piagetian epistemology: equilibration and the teaching of science. Synthese 80:141–162

Russell SJ (1999) Mathematical reasoning in the elementary grades. In: Stiff LV (ed) Developing mathematical reasoning in grades K-12. National Council of Teachers of Mathematics, Reston, VA

Sanni R (2008a) Mathematics task analysis: the evolution of an analytical framework. In: Polaki MV, Mokuku T, Nyabanyaba T (eds) Proceedings of the sixteenth annual congress of the Southern African association for research in mathematics, science and technology education (SAARMSTE). Maseru, Lesotho

Sanni R (2008b) Mathematics teachers' task practices and teacher knowledge. Unpublished PhD Thesis, University of the Witwatersrand, Johannesburg

Sasman M, Linchevski L, Olivier A, Liebenberg R (1998) Probing children's thinking in the process of generalisation. Paper presented at the fourth annual congress of the association for mathematics education of South Africa (AMESA), Pietersburg

Schifter D (2001) Learning to see the invisible: what skills and knowledge are needed in order to engage with students' mathematics ideas? In: Wood T, Scott Nelson B, Warfield J (eds) Beyond classical pedagogy: teaching elementary mathematics. Lawrence Erlbaum Associates, Mahwah, NJ, pp 109–134

Schoenfeld AH (1985) Mathematical problem solving. Academy Press, Orlando, FL

Schoenfeld AH (1988) When good teaching leads to bad results: the disasters of "well-taught" mathematics courses. Educ Psychol 23(2):145–166

Schoenfeld AH (2002) Making mathematics work for all children: issues of standards, testing, and equity. Educ Res 31(1):13–25

Secada W (1992) Race, ethnicity, social class, language and achievement in mathematics. In: Grouws DA (ed) Handbook of research on mathematics teaching and learning. MacMillan, New York, pp 623–660

Sfard A (1998) On two metaphors for learning and the dangers of choosing just one. Educ Res 27(2):4–13

Sfard A (2001) There is more to discourse than meets the ears: looking at thinking as communicating to learn more about mathematical learning. Educ Stud Math 46:13–57

Shamim F (1996) Learner resistance to innovation in classroom methodology. In: Hywel C (ed) Society and the language classroom. Cambridge University Press, Cambridge, pp 105–121

Shavelson RJ, Li M, Ruiz-Primo MA, Ayala CC (2002) Evaluating new approaches to assessing learning. Paper presented at the joint Northumbria/EARLI assessment conference, University of Northumbria, Newcastle

Sherin MG (2002) A balancing act: developing a discourse community in a mathematics classroom. J Math Teach Educ 5:205–233

Simon MA (1996) Beyond inductive and deductive reasoning: the search for a sense of knowing. Educ Stud Math 30:197–210

Sinclair JM, Coulthard RM (1975) Towards an analysis of discourse: the English used by teachers and pupils. Oxford University Press, London

Skemp R (1976) Relational understanding and instrumental understanding. Math Teach 77:20–26

Slonimsky L, Brodie K (2006) Teacher learning: development in and with social context. South Afr Rev Educ 12(1):45–62

Smith JP, DiSessa AA, Roschelle J (1993) Misconceptions reconceived: a constructivist analysis of knowledge in transition. J Learn Sci 3(2):115–163

Staples M (2004) Developing a community of collaborative learners: reconfiguring roles, relationships, and practices in a high school mathematics classroom. Unpublished Phd Dissertation, Stanford University, Stanford, CA

Staples M (2007) Supporting whole-class collaborative learning in a secondary mathematics classroom. Cogn Instr 25(2):1–57

Steen LA (1999) Twenty questions about mathematical reasoning. In: Stiff LV (ed) Developing mathematical reasoning in grades K-12. National Council of Teachers of Mathematics, Reston, VA

Stein MK, Grover BW, Henningsen MA (1996) Building student capacity for mathematical thinking and reasoning: an analysis of mathematical tasks used in reform classrooms. Am Educ Res J 33(2):455–488

Stein MK, Smith MS, Henningsen MA, Silver EA (2000) Implementing standards-based mathematics instruction: a casebook for professional development. Teachers College Press, New York

Sternberg RJ (1999) The nature of mathematical reasoning. In: Stiff LV (ed) Developing mathematical reasoning in grades K-12. National Council of Teachers of Mathematics, Reston, VA

Sternberg RJ, Torff B, Gigorenko EL (1998) Teaching tiarchically improves school achievement. J Educ Psychol 90(3):374–384

Stigler JW, Hiebert J (1999) The teaching gap: best ideas from the world's teachers for improving education in the classroom. Free Press, New York

Swan M (2001) Dealing with misconceptions in mathematics. In: Gates P (ed) Issues in teaching mathematics. Falmer Press, London

Tabulawa R (1998) Teachers' perspectives on classroom practice in Botswana: implications for pedagogical change. Int J Qual Stud Educ 11(2):249–268

Tatto M (1999) Improving teacher education in rural Mexico: the challenges and tensions of constructivist reform. Teach Teach Educ 15:15–35

Taylor N (1999) Curriculum 2005: finding a balance between school and everyday knowledges. In: Taylor N, Vinjevold P (eds) Getting learning right: report of the president's education initiative research project. Joint Education Trust, Johannesburg, pp 105–130

Taylor N, Vinjevold P (eds) (1999) Getting learning right: report of the president's education initiative research project. Joint Education Trust, Johannesburg

Taylor N, Muller J, Vinjevold P (2003) Getting schools working: research and systemic school reform in South Africa. Pearson Education, Cape Town

Triandafillidis TA, Potari D (2005) Integrating different representational media in geometry classrooms. In: Chronaki A, Christiansen IM (eds) Challenging perspectives on mathematics classroom communication. Information Age Publishing, Greenwich, CT

Open University (1997) Course MS221 exploring mathematics, block D: proof and reasoning. Open University Press, Milton Keynes

Volmink J (1990) The nature and role of proof in mathematics education. Pythagoras 23:7–10

Vygotsky LS (1978) Mind in society: the development of higher psychological processes. Harvard University Press, Cambridge, MA

Vygotsky LS (1986) Thought and language. MIT Press, Cambridge, MA

Wallach T, Even R (2005) Hearing students: the complexity of understanding what they are saying, showing, and doing. J Math Teach Educ 8:393–417

Watson A, Mason J (1998) Questions and prompts for mathematical thinking. Association of Teachers of Mathematics, Derby

Wells G (1999) Dialogic inquiry: toward a sociocultural practice and theory of education. Cambridge University Press, Cambridge

Wenger E (1998) Communities of practice: learning, meaning and identity. Cambridge University Press, Cambridge

Wertsch JV (1984) The zone of proximal development: some conceptual issues. In: Rogoff B, Wertsch JV (eds) Children's learning in the "zone of proximal development". Jossey Bass, San Francisco, pp 7–18

Willis P (1977) Learning to labour. Saxon House, Farnborough

Wood T (1994) Patterns of interaction and the culture of mathematics classrooms. In: Lerman S (ed) Cultural perspectives on the mathematics classroom. Kluwer, Dordrecht, pp 149–168

Wood T, Cobb P, Yackel E (1992) Investigating learning mathematics in school classrooms: interweaving perspectives. Paper presented at the ICME 7, Quebec, Canada

Yackel E, Cobb P (1996) Sociomathematical norms, argumentation and autonomy in mathematics classrooms. J Res Math Educ 27:458–477

Index